PATENTS
DEMYSTIFIED

PATENTS DEMYSTIFIED

AN INSIDER'S GUIDE TO PROTECTING IDEAS AND INVENTIONS

DYLAN O. ADAMS

AMERICAN BAR ASSOCIATION
Defending Liberty
Pursuing Justice

Cover design by Kaitlyn Bitner/ABA Publishing.
Interior design by Betsy Kulak/ABA Publishing

Printed in the United States of America.

19 18 17 16 15   5 4 3 2 1

Library of Congress Cataloging-in-Publication Data

Adams, Dylan O., author.
  Patents demystified : an insider's guide to protecting ideas and inventions /
Dylan O. Adams.
    pages cm
  Includes bibliographical references and index.
  ISBN 978-1-63425-167-9 (alk. paper)
  1. Patent laws and legislation—United States. I. Title.
  KF3114.A73 2015
  346.7304'86—dc23
                        2015034390

Discounts are available for books ordered in bulk. Special consideration is given to state bars, CLE programs, and other bar-related organizations. Inquire at Book Publishing, ABA Publishing, American Bar Association, 321 N. Clark Street, Chicago, Illinois 60654-7598.

www.ShopABA.org

Contents

Acknowledgments

Special thanks to my wife, Jen, for her unwavering support of me and this project.

Thanks to Brian Considine as well as Dawn, Doug, and Jen Adams for graciously volunteering their time to edit and review my manuscript.

Thanks to the attorneys and staff at the firms of Davis Wright Tremaine LLP, Orrick Herrington & Sutcliffe LLP, Graybeal Jackson LLP, the Axios Law Group LLP, and RuttlerAdams LLP for their roles in guiding and mentoring my understanding of patents at various stages of my career.

Thanks to my literary agent, Rita Rosenkranz, for making this book possible.

And last, but certainly not least, thanks to the many clients, students, entrepreneurs, innovators, investors, coders, makers, executives, engineers, scientists, techies, and tinkers who inspired this book.

Disclaimer

This book is intended to provide information about patents and intellectual property but should *not* be construed as being legal advice. In fact, a core message of the book is that inventors and companies should seek out a trusted attorney as soon as possible so that intellectual property rights are not inadvertently lost or otherwise compromised. Because the facts surrounding each invention and company are unique, the advice of a competent attorney is necessary to craft an intellectual property strategy that is specially tailored to each individual case.

Moreover, intellectual property laws are always in flux and the information provided in this book may, therefore, be incomplete or incorrect if the laws change. Always consult an attorney to confirm that you are operating under the current state of the law.

Accordingly, the author, and any law firm or company that the author may be associated with, will not be responsible for any negative consequences, damages, or loss of intellectual property rights that might occur by relying on any information provided in this book. Please use the content of this book for inspiration in planning a legal strategy in collaboration with your attorney.

Introduction

I am a patent attorney with a master's in electrical engineering and a bachelor's in biochemistry, and I assist my clients with fascinating and cutting-edge technology on a daily basis. From industry leaders to garage inventors, I work with a full spectrum of clients and a wide variety of technology. Detection crystals for supercolliders, drone technology, military hardware, and even thong underwear for dogs are a few of the inventions with which I have worked. My passion, however, is working with startups and emerging companies. I love the energy and enthusiasm of budding innovators hustling to bring their products to market and I find that these clients are the most in need of reliable counsel on how to protect their ideas and inventions.

Some of these companies may consist of only a few core founders who are working to develop a first product with their own savings as startup capital. Others have raised money through Kickstarter, angel investors, or venture capital funds, but are still pushing to get a first product on the market. Despite being new to the process, these innovators are keenly aware that patents are usually one of the most valuable assets their companies will own and are eager to learn more about the process.

Unfortunately, these same companies typically have a hard time getting good patent guidance. Although general information about the patent process is available from a wide range of resources, those resources do not provide necessary detail or offer a one-size-fits-all strategy that fails to account for the unique needs of each business and product. Seasoned patent attorneys are the only reliable source of a custom-crafted patent strategy that maximizes patent protection and brings value to a company, but attorney time is extremely expensive. Sophisticated and well-capitalized companies already know how to grow and leverage their patent assets and will invest exorbitant sums on a carefully planned and cultivated patent portfolio. Such companies are typically well versed in the nuances of patent strategy or have the budget to hire experts who can provide that expertise. On the other hand, emerging companies are often completely new to the world of patents and are in desperate need of mentorship from a seasoned patent attorney, while also being concerned about the expense involved. Although developing a targeted patent strategy is absolutely crucial at these early stages, new companies tend to be the most under-served and misinformed, and often unknowingly make mistakes that cripple the company before it gets off the ground. From waiting too long and forfeiting all patent rights, to over- or under-spending on worthless patent protection, I see far too many great ideas, products, and businesses that have been compromised beyond salvation and needlessly lost.

This book provides a cost-effective solution to this problem by providing an easy-to-understand insider guide to patents. Based on my first-hand experience with both successful and failed companies of all sizes, I can assist you in learning how to maximize patent protection on any budget, with strategies that can be tailored to companies with any business plan or product. Instead of being intimidated and confused by patents, you will learn how to proactively work with a patent attorney to craft a customized patent plan. This book will teach you insider patent strategies that large successful companies already know and that many serial entrepreneurs learn only after a few failed attempts. Whether you are building up the courage to start your own company, preparing to launch your flagship product, or already have a patent application filed, this book will give you crucial insight into planning and growing a patent portfolio at any stage. Most importantly, this book will help you to understand patents and the patent process so that you can effectively plan an intellectual property strategy that maximizes protection within a reasonable budget.

This book is for tinkers, entrepreneurs, and innovators, as well as businesses with products or processes that could be patented. It is useful to inventors and companies at any stage of development, and the content is even relevant where patent applications have already been filed or where patents have already been issued. The content is also applicable to all sorts of products and technology, including simple mechanical products, computer software, electronic devices, biotechnology, production methods, business methods, and improvements on existing products or processes. Additionally, although the focus is on U.S. patents, foreign companies seeking U.S. patents will also benefit from this book.

We begin by discussing the four main types of intellectual property (trademarks, copyrights, trade secrets, and patents) so that you can understand what types of inventions can be protected by a patent and what innovations are better protected in other ways. From there, we walk through all the stages of the patent process—from protecting yourself during invention and product development to selecting a patent attorney and working with him or her to draft patent applications and working through the examination process at the U.S. Patent and Trademark Office. Finally, we discuss advanced techniques for building a patent portfolio having multiple patents and offer advice on how to license, enforce, and otherwise leverage patent assets.

It's likely that you are reading this book because you already have an inkling of how powerful patents can be. You know that growing companies can be propelled to success by patenting ideas and inventions. The information in this book will enable you to avoid years of trial-and-error or the high cost of a patent attorney providing you with a personal master class on the patent process. Welcome to the insider world of patents. Let's get started.

4 TYPES of IP

1.) TRADEMARK
2.) COPYWRITE
3.) PATENTS
4.) TRADE SECRETS

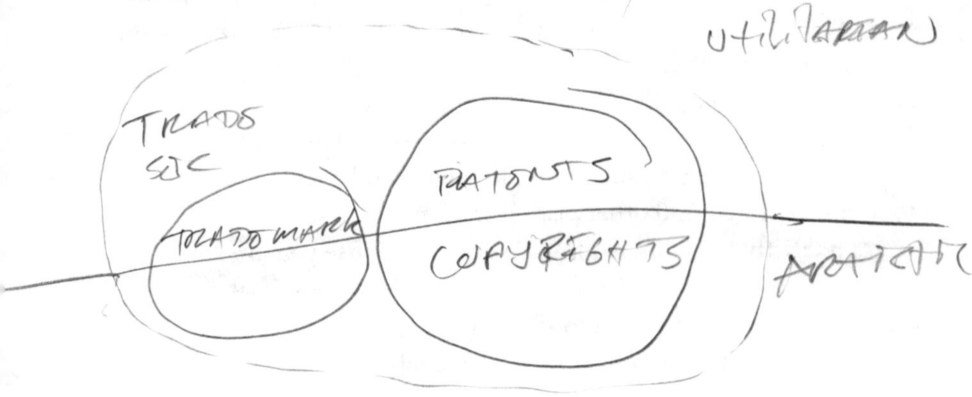

utilitarian

TRADE
SEC
TRADEMARK

PATENTS
COPYRIGHTS

artistic

Types of Intellectual Property

"The Congress shall have Power To . . . promote the Progress of Science and useful Arts, by securing for limited Times to Authors and Inventors the exclusive Right to their respective Writings and Discoveries."

—United States Constitution, Article 1, Section 8, Clause 8

Intellectual property (IP) protects creations of the mind and patents are just one type of IP. To understand how patents protect your ideas and inventions, it is important to also understand what patents *cannot* protect and where other IP rights might be applied instead. In some cases, several forms of IP can protect the same product or invention in different ways, but in other cases current IP laws do not provide much protection. This chapter compares and contrasts patents, trademarks, copyrights, and trade secrets—the four types of IP.

Although this chapter provides a solid framework for determining how the four types of IP may apply to your inventions, use the information as a starting point for getting expert advice on developing a holistic IP strategy for your business. In the end, because each company has its own unique business plan, technology, and budget, you need a plan that is tailored to your needs. Fortunately, once established, many portions of your IP protection strategy can be implemented without the assistance of an attorney if you do not yet have a robust IP budget. In fact, as discussed later in this chapter, there are several simple and absolutely free techniques that can significantly raise your IP profile.

OVERVIEW OF THE FOUR TYPES OF INTELLECTUAL PROPERTY

When considering intellectual property, it helps to first divide the world of inventions into artistic and utilitarian things. Artistic inventions include books, articles, movies, musical compositions, and photographs. Utilitarian inventions include mechanical devices, computer hardware, biotechnology, and manufacturing methods. By definition, patents protect utilitarian inventions and copyrights protect artistic inventions. Trademarks and trade secrets can protect both useful and artistic inventions.

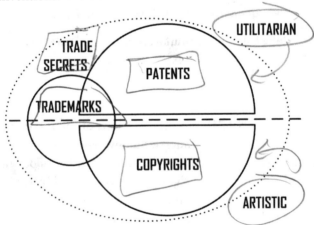

FIGURE 1.1   The Logical Domains of IP

As shown in Figure 1.1, patents and copyrights are diametrically opposite forms of protection that apply to useful and artistic works respectively—there is no overlap. Trade secrets and trademarks might

*Code protoction as
is protoction work
Literary*

protect some of the same inventions that patents and copyrights do, but not the same types of inventions. Trademark law applies to "marks" used publicly in commerce to indicate the origin of goods or services. In contrast, trade secret law provides protection to inventions or assets that are held in secret and not available to the public. The overlap between trade secrets and the other types of IP has an important conflict because at some point obtaining other IP protection makes these inventions public and therefore negates any protection as a secret. As discussed in more detail below, trade secret law can initially protect innovations that are later protected by patents, trademarks, or copyrights, but at some point inventors must choose between trade secret protection and the other protection options.

Although the four types of IP protect different inventions, this is not to say that a given product will be limited to one type of IP protection. In fact, most products actually include tens if not hundreds of unique inventions that are each eligible for one or more forms of IP protection. One of the most important takeaways from this book should be that you view innovations that you once considered as a single invention to actually be an amalgamation of many protectable inventions.

For example, let's say you invent a smartphone app that identifies faces and then speaks the names of the people who are identified. The code that makes up the app is considered a literary work and is protectable by copyright. On the other hand, as discussed in more detail in the coming chapters, the functionality of the app may be protectable by patents in several different ways—each being defined as a unique invention. Instead of being a single invention, this relatively simple app could actually be the subject of at least one copyright and several patents.

Accordingly, as discussed in more detail in the coming chapters, creating a comprehensive IP strategy for your business and products starts with "mining" for protectable inventions and determining which ones are most valuable given your business plan, competitors, and potential customers. Relative cost and value of securing protection can then be used to determine where to focus your IP budget. However, before embarking on your mining expedition, you will need to know what you are looking for.

TRADEMARKS

Federal law defines a trademark as a "word, name, symbol, device or any combination thereof, which is used to distinguish goods of one person from goods manufactured or sold by others, and to indicate the source of the goods, even if the source is unknown."[1] Simply put, trademark law protects identifying "marks" used in association with goods and services. The most common trademarks are brand or product names, logos, slogans, and combinations of such marks. For example, several Nike trademarks are shown together in Figure 1.2. The word 'NIKE' alone is a trademark and the word 'NIKE' in distinctive font is another trademark. The classic Nike swoosh graphic and the slogan "JUST DO IT" are also individual trademarks. Additionally, any combination of these individual marks would also be considered a unique trademark on its own as in the combined graphic shown.

FIGURE 1.2

Trademarks are not limited to words and designs. Sounds, colors, smells, textures, shapes, motions, and the appearance of products, packaging, or places of business can also be protectable by trademark law. For example, Owens Corning has a trademark on the color pink in relation to fiberglass insulation. Metro-Goldwyn-Mayer (MGM) has a trademark on the sound of a roaring lion that is part of the company graphic it presents at the beginning of MGM films.

---

1. Lanham Act §45.

FIGURE 1.3   Example of an MGM Trademark

However, the protection afforded by trademark law is not universal and only provides exclusivity in uses of the protected mark in relation to certain goods or services. For example, the Owens Corning trademark on the color pink does not enable the company to control all uses of the color pink throughout the universe—it cannot even control uses of the color pink on all commercial products. The protection only gives Owens Corning exclusivity in making pink fiberglass insulation and possibly some closely related products or services.

The true scope of protection afforded by trademark law is largely misunderstood, and is exemplified by a famous catch-phrase trademark that I still frequently get calls about. In 2004, Donald Trump's "The Apprentice" was one of the hottest shows on television. The show features a group of people, sometimes celebrities, who compete in business challenges and are eliminated one by one until a single winning contestant remains. At the end of each program, a contestant is eliminated in a boardroom by Donald Trump, who exclaims "You're fired!" This simple catch-phrase became wildly popular and quickly grew into a central brand for the show and for Trump himself. When the media began to report that Trump filed a trademark for the phrase "You're fired!" public reaction was one of outrage and disgust for the trademark system. People largely believed that Trump's trademark would allow him to sue anyone who said "You're fired!" and require anyone who used the phrase to pay him a royalty.

What most media reports failed to mention or clearly explain was that trademark law only provides protection for uses of a mark in relation to specific goods and services that the applicant actually sells or intends to sell. Trump's trademark applications sought to cover the use of the "You're fired!" mark on products like paper goods, home furnishings, pillows, housewares, linens, toys, and sporting goods. Although Trump ultimately abandoned all of his trademark applications, several registered trademarks for the phrase "You're fired!" are alive and well. For example, Mark Burnett, producer of "The Apprentice," and other shows, such as "Survivor," and "The Voice," owns several trademarks for "You're fired!" claiming "entertainment services in the nature of a reality television series" and for goods such as "shirts, hats, and beverage glassware."[1] Another completely unrelated and independently owned "You're fired!" registered trademark claims services of a do-it-yourself pottery business.[2]

The idea of universally protected catch-phrases still persists and I regularly confer with people who want to collect royalties from anyone who uses a new saying or slogan they believe they have created. These people are disappointed to hear that trademarks offer protection in relation to specific goods and services and, most important, protect only those goods and services that a company actually sells or intends to sell. Trademark applications can be filed based on an intent to use the mark in commerce, but eventually applicants must prove that they are using the mark in association with each and every claimed good or service. If you do not sell or intend to sell goods or services, you cannot claim the benefits of trademark protection.

This requirement of using the mark in commerce in relation to goods or services makes sense when considering that the purpose of trademark protection is to promote economic efficiency by giving consumers confidence in the source of goods or services when making buying decisions. For example, when you step into a Starbucks or McDonalds you are guaranteed one thing—consistency. Regardless of whether you are in Seattle, San Francisco, or Manhattan you know that coffee at a Starbucks café will taste essentially the same and burgers at a McDonalds will not vary from place to place. Without trademark protection, competing business could hijack the brand and goodwill

---

1. U.S. Trademarks 3263242, 3469268, 3269040 to JMBP, Inc.
2. U.S. Trademark 3208352 to You're Fired LLC for: providing studio facilities for paint your own pottery and create your own mosaics; providing do-it-yourself ceramic studio facilities with precast ceramic pottery that is painted on-site by customers and then glazed and/or kiln-fired; and providing do-it-yourself studio facilities with mosaic tiles that are pieced together to create mosaics on-site by customers.

associated with the Starbucks or McDonalds name and steal customers that would have chosen a legitimate store.

The key is confusion. Trademark law seeks to prevent consumer confusion by assuring that they know who made a given product when making buying decisions. On the other hand, lack of confusion allows the same exact trademark to be used by many companies simultaneously. For example, a search for "EAGLE" at the United States Patent & Trademark Office (USPTO) returns nearly 1,700 active registered trademarks with the vast majority of them being owned by completely separate companies or individuals. These trademarks can be used at the same time because consumers are not likely to be confused between any given mark. Consider that the goods and services of these "EAGLE" marks range from potato chips, to shirts, to insurance. If buyers pick up a bag of Eagle brand potato chips, they are not likely to assume that the chips were produced by Eagle shirt company or Eagle insurance company. Each of these "EAGLE" marks is equally viable because the goods and services are not overlapping to the point that consumer confusion will occur.

TRADE SECRETS

In stark contrast to trademarks that are used publicly in commerce, trade secret laws protect certain inventions and assets that a business decides to keep secret. Unlike patents, copyrights, and trademarks, there is no official registration process for trade secrets and the laws related to trade secret protection may be different from state to state. By simply keeping certain things secret in the right way, a company can be afforded the protections of trade secret laws.

Despite some jurisdictional differences, trade secret laws almost universally require protectable trade secrets to (1) not be generally known to the public; (2) have economic value derived from being nonpublic; and (3) be the subject of reasonable efforts to maintain their secrecy. Not being generally known to the public means that the information or technology should not be available from the party trying to keep it secret, nor should it be available from any other source. However, a novel combination of known information or technologies can still be regarded as nonpublic. For example, the individual elements of customer lists may be known, but if the compilation of elements is not generally known, such proprietary lists can still be the subject of trade secret protection.

A company must also make reasonable efforts to keep its trade secrets nonpublic. Whether secrecy efforts are considered reasonable varies by state and from case to case. However, some basic rules for protecting trade secrets include clearly marking materials as being secret; requiring personnel handling them to sign nondisclosure or noncompete agreements; having computers with secret information password-protected; and keeping secret items in locked rooms or locations that are not accessible by unauthorized persons.

Public companies in the United States are estimated to own more than $5 trillion in trade secret information,[1] which can include both useful and artistic inventions that may or may not be ultimately protectable by patents, trademarks, or copyrights or that may remain exclusively a trade secret. Examples of innovations that typically remain trade secrets include production methods, formulas, client lists, business plans, financial records, and positive or negative know-how. Some of the most valuable trade secrets are embodied in products that you likely come in contact with on a regular basis.

For example, the formula for making Coca-Cola has been a closely guarded secret for more than 125 years, since it was invented by Dr. John S. Pemberton in 1886. The formula was initially shared exclusively with a small core group within the business and was purportedly written down for the first time in 1919 when the company was bought by Ernest Woodruff and a group of investors. This single written copy was housed in the same Atlanta bank vault from 1925 to 2012, when it was transferred to a new state-of-the-art vault at the World of Coca-Cola museum. Other famous product formulas protected by trade secrets include WD-40 lubricant, Listerine mouthwash, Bush's Baked Beans, KFC chicken, and even the method of determining the *New York Times* best sellers list.

Trade secret protection is also helpful for protecting innovations that will later be the subject of patent, trademark, or copyright applica-

---

1. *See* John P. Hutchins, *The Corporation's Valuable Assets: IP Rights under Sarbanes-Oxley*, in 26th Annual Institute on Computer & Internet Law 289, 291-292 (PLI Intellectual Property, Course Handbook Series No. G-859, 2006).

tions. For example, in the initial stages of developing patentable technology and while the invention remains a secret within the company, trade secret protection can fill the gap until patent applications are filed, and will remain a viable protection option until the patent application is published and becomes publicly available.

Alternatively, in some cases, even though an invention would be eligible for patent protection, a business may decide to forego that protection in favor of trade secret protection. Patents will have a term of around 20 years before the invention becomes public domain, whereas trade secret protection can last indefinitely. However, keeping something a trade secret for too long can cause the forfeiture of patent rights and, if the secret subsequently becomes exposed or otherwise known to the public, then all protection is lost.

COPYRIGHTS

Copyrights protect "original works of authorship fixed in any tangible medium of expression" as defined by federal law.[1] Such works include literary works; musical works; dramatic works; pantomimes and choreographic works; pictorial, graphic, and sculptural works; motion pictures and other audiovisual works; sound recordings; and architectural works.[2] This covers a wide variety of artistic and expressive works, including books, blog posts, movies, songs, paintings, and even the code for software.

However, an important limitation is that copyrights only protect expression and not an underlying idea, product, or invention that is described or shown in the work of authorship and will not protect useful products or articles.[3] As illustrated by Figure 1.1, this puts copyrights in direct opposition to patents, which exclusively protect utilitarian articles and products. Failing to completely understand the respective coverage of patents and copyrights often leads to an inadvertent forfeiture of patent rights or incomplete protection of a product or invention.

---

1. Copyright Act § 102(a).
2. Copyright Act § 102(a).
3. Copyright Act § 102(b) "In no case does copyright protection for an original work of authorship extend to any idea, procedure, process, system, method of operation, concept, principle, or discovery, regardless of the form in which it is described, explained, illustrated, or embodied in such work."

The tricky and confusing protection of products such as furniture and clothing is a great illustration of how patents and copyrights protect the same product differently, yet synergistically. Although the design of furniture is often aesthetically pleasing and artistic, such pieces are nonetheless considered to be utilitarian because of their intrinsic function as a place to sit or lounge. Accordingly, the physical design and configuration of furniture cannot be protected by copyright and must instead be protected by patents. At the same time, copyrights can protect important aspects of furniture where patents fall short. Fabric designs or sculptural works are non-utilitarian works of authorship that are protected by copyright, and these works do not lose their protection when applied to useful articles. Accordingly, the pattern of fabric on a couch or a carving on a wooden chair would be protected by copyright, but the overall physical design and configuration of the furniture would still need to be protected by a patent.

FIGURE 1.4   Design Patent for a Couch

Unfortunately, I routinely see misunderstanding these subtle differences in patent and copyright protection result in the complete loss of important patent rights. For example, after several years of selling a line of clothing that included some novel design features, a potential client approached me to discuss asserting her copyrights against a competitor that was producing similar clothing that had these same novel features. She had self-filed a series of copyright applications that included numerous photographs and descriptions of the novel features in her clothing line. Needless to say, she was quite disappointed when I informed her that these copyrights protected nothing more than the photographs that she filed and possibly the specific wording that described her invention. The novel design of her clothing was not protected by copyright because clothing is considered a utilitarian article and because copyrights cannot protect underlying ideas or concepts. Moreover, because the product line had been on sale for a number of years, the time limit to patent the novel design features had expired and her important invention was now effectively in the public domain and usable by anyone. She therefore had no ability to exclude others from using her invention and could not require them to pay a license to use the design. Her competitors were able to freely make and sell her novel design with impunity and there was no way to recover the rights she had unknowingly forfeited.

INSIDER TIP: MARKING YOUR IP

You may be aware of markings or symbols such as "patent pending," TM, ©, and ®, that mark products, logos, or names, but the meaning and proper use of these identifiers is not always clear. Many new businesses are surprised to find that they can even use some of these IP markings without any formal registration or paying fees. For example, both the TM and ® are used to identify trademarks, but ® can only be used after federal trademark registration. However, the TM symbol can be used freely regardless of registration. Markings like "pat. pend." can only be used once a formal patent application has been filed for a product. On the other hand, the © symbol can be used to identify copyright protection regardless of federal registration. Tasteful use of the TM and © marks is therefore a great way to market and identify your valuable IP for free before you pay a cent in fees for formal registration.

Copyrights can form an important part of an intellectual property portfolio, but understanding their proper use and complement to other forms of protection is essential when planning your personalized intellectual property protection strategy.

PATENTS

In contrast to copyrights, patents can protect an underlying form or function of utilitarian inventions. Utility patents protect any new and useful process, machine, manufacture, or composition of matter, or any new and useful improvement thereof.[1] Mechanical devices, computer software and hardware, biotechnology, pharmaceuticals, and methods of making or using these inventions are just a few examples of the wide variety of subject matter that can enjoy patent protection. A second type of patent—the design patent—protects any new, original, and ornamental design for an article of manufacture.[2] Much like copyrights protect the ornamental design of sculpture and other artistic works, design patents protect the ornamental design of utilitarian products without protecting the underlying concept of such products.

UTILITY VS. DESIGN PATENTS

Utility Patents	Design Patents
Cover idea or product generally	Cover one specific design
Substantially broader protection	More narrow protection
More expensive	Substantially less expensive
Takes 2–5+ years to issue	Can issue in 1 year or less
Much more difficult to obtain	Almost always allowed

Although design patents typically represent just 5% of total patent applications filed each year and are only useful for protecting certain types of products, design patent protection has become an increasingly valuable, yet often overlooked IP tool. Recent court cases have greatly expanded the scope of protection that design patents afford. Despite protecting only a single specific design shown in the drawings of the

---

1. 35 U.S.C. §101.
2. 35 U.S.C. §171.

[handwritten annotations: DESIGN PATENTS ARE VERY SPECIFIC / Deviate even slightly & you can get past it]

patent, a design patent is substantially less expensive because of the simplicity of the application and high rate of application allowance. Moreover, the USPTO backlog for examining design patents is much less compared to utility patents, so design patents can be issued in a matter of months instead of years.

UTILITY PATENT CLAIM:

1. A tool comprising a hammer-head, wrench slot, and first nail-puller at one end of an extended shaft, and a second nail-puller at a second end of the shaft.

DESIGN PATENT CLAIM:

1. The ornamental design for a tool as shown.

In contrast, utility patents can provide substantially broader protection because the invention protection is defined by a claim of words instead of drawings. In the example shown above, the design patent drawing claim protects one specific design of a tool having two-nail pullers, a hammer-head, and a wrench slot. By changing the proportion, shape, or configuration to a design that looks different from the picture, a competitor would be able to circumvent and not infringe the design patent. The example utility patent claim provides significantly broader protection because, as long as a competing product has all of the elements described, the product will still infringe the utility patent. Unlike the design patent, the utility patent claim covers the product shown in the design patent along with numerous other designs that may be substantially different in proportion, shape, or configuration.

To be eligible for patent protection, a claimed invention must be adequately described in the patent application and must be new and non-obvious when compared to existing products or technology disclosures (known as "prior art"). Adequate disclosure requires that a person having ordinary skill in the art would be able to make and use the invention based on the description provided.[1] In other words, an average person who works in the relevant area of technology should have enough information from the patent alone to construct and operate the claimed invention without undue experimentation.

A claimed invention is considered new when no single piece of prior art describes the invention in full. In contrast, an invention is non-obvious when that same imaginary person having ordinary skill in the art would not find the invention obvious in view of one or more prior art references. The vast majority of patent applications that are rejected and never issue as an enforceable patent ultimately fail because they are deemed obvious by a patent examiner at the USPTO. Because understanding this important standard is essential in evaluating the patentability of an invention, when drafting a patent application, and during the examination process, the coming chapters will delve deeper into the concept of obviousness and novelty. Additionally, because utility patents are more complex and add greater relative value to a patent portfolio, the remainder of the chapters focuses on utility patents, but do not forget that design patents can have an important role in a patent portfolio and an overall intellectual property protection strategy.

BUILDING AN INTELLECTUAL PROPERTY PORTFOLIO AND PROTECTION PLAN

Even a small business can produce products and innovations that could conceivably be protected by tens, if not hundreds, of patents, trademarks, and/or copyrights. With an unlimited budget, it would be

---

1. 35 U.S.C. § 112.

possible to fully protect each of the multitude of unique inventions that are inherently embodied in even the simplest of products and innovations. In the real world, however, businesses must carefully develop an IP strategy that extracts the greatest protection and marketing value within a reasonable budget. A difficulty that most innovators face is identifying all the possible ways that a given product or technology can be protected, putting a relative value on the array of options, and coming up with an IP budget that is sufficient yet not wasteful. Unfortunately, there is no cookie-cutter solution because each business is unique and good advice can be hard to come by.

This chapter provides insight on how to protect IP with patents, trademarks, copyrights, and trade secrets, but it should not be a substitute for the expert evaluation and planning of a skilled patent or IP attorney. In the beginning stages of developing an idea or business, a short amount of time with a trusted attorney is cheap (and often free) insurance against inadvertently overlooking or forfeiting IP rights that would be an essential foundation for an emerging company.

To illustrate how an IP evaluation can uncover numerous inventions hidden within a single product or business, let's walk through an evaluation of a theoretical growing business that currently only has a single product—a tractor.

FIGURE 1.5

Because tractors are utilitarian products, patent protection should be the first option that comes to mind. Utility patents could protect the overall structure or function of the tractor or may protect any separate part or collection of parts. Individual utility patents could cover the exhaust system, transmission, engine, drive train, tires, electronic systems, or any part that is sufficiently new and non-obvious over existing parts. Utility patents could also protect proprietary methods that relate to the operation of any of these parts, either individually or collectively, including methods of manufacturing these parts. Again, any such method simply needs to be new and non-obvious over known methods.

Numerous design patents can also protect this single model of tractor. In addition to protecting the look of the complete vehicle, design patents can protect ornamental aspects of any specific part. The cab, tire treads, engine cover, fenders, and even the ornamental design of internal components could be the subjects of design patents. Design patents on the overall design and also on the design of individual parts can be important because of the different scope of protection that each provides. For example, a competitor could work-around a design patent on the whole tractor by making changes to some parts of the protected design and leaving some parts exactly the same. In contrast, design patents on specific parts provide protection regardless of the look of the parts adjacent to the protected design or even whether the protected parts are attached to a tractor or other vehicle.

Because tens, if not hundreds, of design and utility patents could protect this single product, the company would need to put a priority on patents that protect the most valuable designs, functions, configurations, or methods. As discussed in the coming chapters, choosing how to build a patent portfolio around a given product also depends on the business strategy of a company. Along with providing exclusivity in patented technology, a well-crafted patent portfolio can also be used to attract customers, business partners, and investors.

Trademarks can protect both utilitarian and artistic aspects of this specific tractor product and the business as a whole, including a business name, product name, slogans, logos, and even distinctive designs or colors that relate to the tractor or business in general. For example, the John Deere Company is famous for its distinctive yellow and green farm equipment, which is an important color trademark of the business. Its leaping deer logo and the "JOHN DEERE" name shown in Figure 1.6 are also examples of company trademarks. Distinctive parts of our example tractor can also be the subject of trademark protection. If the company considers the look of specific parts to be important to differentiating its brand, trademarks may provide limited exclusivity for these elements. For example, in addition to design patent protection on various body designs, Bentley Motors has trademarks on two distinctive designs of the grilles of Bentley vehicles.[1] Even though trademarks can protect the ornamental design of utilitarian articles as design patents do, trademark protection is more limited than a design patent. Trademarks provide exclusivity in relation to specific goods or services, but design patents apply to an ornamental design of an article regardless of how the article and design are used. Trademark protection can last indefinitely, as long as the mark is used in conjunction with the claimed goods and services, whereas design patent protection only lasts for 17 years from when the patent issues.

FIGURE 1.6

---

1. U.S. Trademarks 1512167 and 1246990 to a grille design; U.S. design patent 686535 to a motor vehicle.

FIGURE 1.7   Design patent for a Bentley automobile

Copyrights can also be important in protecting our example tractor, although less so than patents, trademarks, and trade secrets, which are better suited for utilitarian innovations. Because copyrights protect expressive works, artistic logos or paint designs of the tractor may be protected by copyright. Additionally, marketing materials for the tractor and company, including photographs, commercials, print advertisements, and promotional writings can be protected by copyright. Protection afforded by a copyright can last the lifetime of the creator plus 70 years or more.

Trade secret protection is beneficial as temporary protection or may be used for long-term protection of certain innovations related to the tractor. For example, while developing the features, marks, and parts that may be protected by patent, copyright, or trademark discussed above, the company may use nondisclosure agreements with employees and contractors to control disclosure outside the company. However, once such innovations are disclosed or made available publicly, trade secret protection will be lost. Where innovations related to the tractor cannot be reversed engineered from the product, trade secret protection may remain active as long as secrecy is maintained. For example, methods of producing certain parts, or the tractor as a whole, may remain a trade secret along with proprietary formulations of paints, lubricants, oils, or materials. For company assets such as client or vendor lists, trade secret law may provide protection where no other type of IP protection is applicable.

CHAPTER 1 SUMMARY

- Trademarks protect identifying "marks" (e.g., logos or slogans) used in association with specific goods or services.
- Trade secret law provides remedy for misappropriation of economically valuable knowledge that was not known publicly and where reasonable efforts were used to maintain its secrecy.
- Copyrights protect artistic and expressive works including books, websites, movies, songs, photographs, paintings, and even the code of software.
- Patents protect an underlying form or function of utilitarian inventions (e.g., mechanical devices, computer software and hardware, biotechnology, pharmaceuticals, and methods of making and using such inventions).
- A single product can be the subject of many trademarks, copyrights, trade secrets, and patents.

2

Why Invest in Patent Protection?

"Every patent shall contain . . . a grant to the patentee . . . of the right to exclude others from making, using, offering for sale, or selling the invention throughout the United States or importing the invention into the United States. . ."

—Title 35 of the United States Code, Section 154(a)(1)

Before embarking on the patent process, it is important to understand how your business can derive value from the asset that is created. In addition to the well-known benefits of being able to exclude competitors from copying the patented invention or charging a royalty to use the technology, there are many other important reasons to seek patent protection. For example, a patent portfolio can be one of the most valuable assets of a company in terms of both book value and marketing value. Patents mark a company as an innovator and are prestigious in the eyes of potential customers, investors, and business partners. On the other hand, patents are feared by existing competitors and act as a deterrent for anyone considering entering the market. Value provided by these many benefits often goes unappreciated by companies unfamiliar with patent assets and may result in significant under-investment in a patent portfolio, which can be worth tens if not hundreds of times the cost of acquiring the patents. This chapter begins by introducing the many benefits derived from the basic right to exclude and then examines other important benefits that are often overlooked by patent owners.

THE RIGHT TO EXCLUDE OTHERS

A U.S. patent provides an exclusionary right—the right to exclude others from making, using, offering for sale, or selling the invention throughout the United States or importing the invention into the United States for a limited term. However, patents *do not* provide the right to practice the invention. For example, a pharmaceutical company can invent a new drug for treating cancer and can patent the compound itself and a method for using the drug in treating cancer, but the Food and Drug Administration (FDA) can still prevent the company from selling the drug if the patented treatment is considered to be unsafe. Despite having a patent, these companies still face the arduous task of seeking FDA approval for their treatments. In fact, FDA approval typically occurs in parallel with the patent process.

Patents are only effective within specific countries, so a U.S. patent is enforceable only within the boundaries of the United States. For example, once you have an issued U.S. patent you can stop anyone within the country from making, using, selling, or offering to sell the claimed product. This applies to competitor companies that manufacture or sell the product and even to customers who buy the patented product. Similarly, for patented methods, both companies and individuals can be prevented from performing the method. On the other hand, without patents in foreign countries, holders of U.S. patents are unable to prevent anyone from freely making, using, and selling the patented invention anywhere in the world aside from the United States. However, infringing products can be stopped at ports and borders and denied entry into the country.

The exclusionary rights endowed by issued patents allow owners to exclusively exploit the patented invention or selectively allow others to practice the invention with permission. Known as a patent license, giving such a permission is typically in exchange for a cash payment or a portion of revenue from sales related to the patented invention. Alternatively, the patent can be sold to a new owner with contract terms that can be similar to the terms of a license.

EXCLUSIVITY IN THE MARKETPLACE

One option for leveraging patent assets is preventing all competitors from making, using, or selling a patented invention. Having a legal monopoly on a product can be immensely profitable, given the right market conditions. For example, pharmaceutical companies such as Pfizer have made billions of dollars in profit by excluding other manufacturers from producing and selling the drugs it develops. The exclusivity with respect to drugs is so valuable that it tends to be big news when major drug patents expire. In fact, the phrase "going generic" specifically means that a patent on a name-brand drug is expiring and that other manufacturers are able to begin producing and selling the drug.

During the term of its drug patents, the name-brand company has a monopoly and, as the only source for the drug, can set its price. As a result, the cost of drugs that are still on-patent is exorbitantly high compared to the price of a drug once it goes generic. In the case of Pfizer's cholesterol drug atorvastatin, which has the brand name Lipitor, the price was $4 or more per pill while Pfizer's patent was in force, and has dropped to 50 cents or less per pill since the drug went generic, with some pharmacies even offering it for free as a loss leader promotion. With Lipitor sales of around $10 billion per year, Pfizer's total profit dropped 19 percent immediately after the drug went generic.

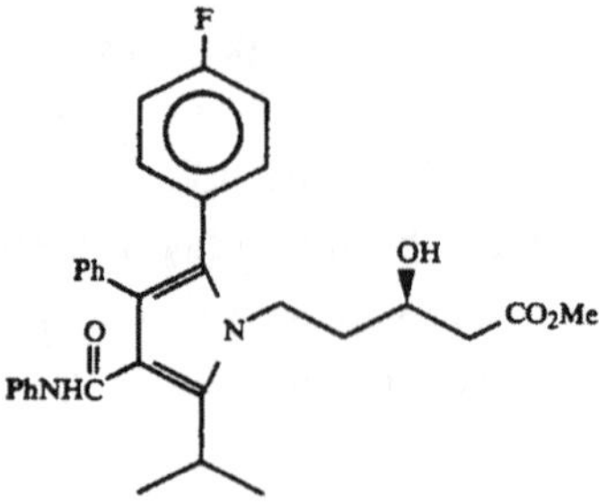

FIGURE 2.1   Chemical Formula from
Pfizer's Lipitor Patent – No. 5,273,995

In a market where demand for a patented product is high, controlling the supply as the only source makes it possible to keep prices high by preventing the market from being flooded with the product. This becomes especially important for products that can be quickly and easily manufactured by competitors or when competitors have superior manufacturing and distribution capabilities.

LICENSING PATENTED TECHNOLOGY

As an alternative to being an exclusive source of the patented product, it is possible to allow other companies to sell, distribute, or otherwise be a source for the technology. Known as a license, such a contract is essentially an agreement not to enforce patent rights against selected parties and can be exclusive or nonexclusive. An exclusive license includes an agreement not to give anyone else a license indefinitely or for a defined period of time. A nonexclusive license maintains the option to license the patent to any number of additional parties. Licenses are typically given in exchange for an ongoing royalty and/ or a cash payment.

Licensing is a great option when a company is unable or does not want to meet the full extent of market demand by itself. For example, if a patent holder only has distribution in one region of the country, it can be beneficial to give a license to competitors in other regions with an agreement to only sell the technology in territories that do not compete with the distribution region of the patent holder. By so doing, the patent holder can receive market exposure and revenue in otherwise unreachable areas.

Patentees can also license their patents when they do not sell a product or otherwise practice the patented technology themselves. In some cases, certain patented technology may not fit into an existing product line so the business may choose to realize value from the technology through licensing revenue instead of as the product producer or seller. In other cases, a company may develop and patent technology with no intention of producing a product itself, or it may buy patents with the intention of profiting exclusively from licensing.

Universities and other educational institutions have traditionally developed technology without the intention of creating businesses around such inventions. Instead, academic researchers typically focus on teaching and advancing their field of study, not on developing products or services that would be profitable to a company. In fact, many academics prefer research without the constraints of a business context so that their focus can be on exploration and discovery instead of on profit. With a focus on teaching and academic research, most universities are not well equipped to produce and market products that are derived from the patented technology. Instead, educational institutions will patent their innovations and license them to companies that can better exploit the technology.

However, universities are increasingly becoming direct incubators for startups where professors and students work together in developing and founding businesses based on technology born in their labs. These companies will eventually spin off into independent businesses and the academic founders often remain heavily involved, with the university being a shareholder and deriving licensing revenue from the patents.

Individual inventors can also profit from developing and licensing technology without having to produce, market, and sell products. In fact, I have several clients who make a living from developing products and licensing the rights to companies that then manufacture and sell the patented products. Ongoing royalties provide these clients with a steady income for years and provide the freedom to invent and license patents full-time. However, these clients are a tiny minority in contrast to the overwhelming majority of individuals who fail to license their

patented technology. The unfortunate truth is that it is extremely difficult to license an idea as an individual inventor.

PATENT PROSECUTION, EXAMINATION, AND LITIGATION

The process of drafting, filing, and working with a patent examiner at the USPTO during the examination process is known as
"patent prosecution." This is not to be confused with "patent
litigation," which refers to the resolution of patent disputes in a
court of law. Patent litigation can include suits related to patent
infringement and whether an issued patent might be invalid for
various reasons. For purposes of clarity, the term "examination" is used instead of "prosecution" throughout the book, but
keep in mind that these terms are interchangeable.

First, most people greatly underestimate the challenge of getting
the attention of executives who have the ability to offer a license, not
to mention the difficulties in getting an opportunity to pitch an idea
to these decision-makers. The majority of companies have a policy of
not responding to unsolicited offers to license, and many will only
work with inventors who are referred or have licensed technology to
the company in the past. Second, most companies are not interested
in licensing bare ideas for products and will only license products that
have a proven market and track record. Licensing theoretical products
is considered to be an unacceptable risk, especially when profit margins
are guaranteed to be smaller because of ongoing royalty obligations.

Also, be wary of invention promotion services that promise to get
your product in front of companies that will want to license your idea.
Although such companies may have better connections and know-how
than the average person, the chances of successfully landing a contract
are still extremely low. The exorbitant upfront fees and low success
rate of these companies make them a poor option for nearly all inven-

tors. While it is possible for individual inventors to make money from licensing their ideas, keep in mind that this business model is more difficult without the right connections or a track record.

On the other hand, established companies can derive significant revenue from a smart licensing program. For example, IBM enjoys revenues of around $1 billion per year from licensing revenues alone, which is fueled by an ongoing strategy of patenting the fruits of its research and development efforts. In 2013 alone, IBM secured 6,500 new patents. Coupled with an aggressive strategy of obtaining licenses from competitors and partners alike, IBM derives a significant portion of its revenue from its patent portfolio.

Another licensing strategy is to simply own patents and license technology without ever producing products or providing services. Known as non-practicing entities (NPEs) or "patent trolls," these companies make money almost exclusively from licensing revenue. Some are shell companies that are created for the sole purpose of holding and enforcing single patents or a small group of patent assets. Filing patent lawsuits against large groups of potential licensees is how such companies typically open licensing negotiations. On the other hand, other so-called patent trolls selectively buy large pools of patent assets in specific technology areas and will even file their own patents to expand and fill in the gaps of these patent portfolios. Although the actual revenue of these large patent trolls is a closely guarded secret, many speculate that these companies quietly rake in billions in licensing revenue every year. Despite being a highly controversial and ethically debatable business practice, few will argue that being a patent troll can be extremely profitable.

Regardless of the size of the company or whether it even practices its patented technology, patent licensing can be a significant source of revenue for a business. Building a patent portfolio may be a sizable investment but, if properly monetized, it can provide a solid income stream that far exceeds the original cost of securing patent rights.

SALE OF PATENTS

Just like tangible business assets, the intellectual property of a patent can be bought and sold. In contrast to licensing where a company maintains ownership of a patent and gives others permission to practice or use the patented technology, selling a patent transfers all rights to the new owner. Accordingly, a new owner can exclude others and offer licenses just as the original owner was able to do. Patents are most often sold for a one-time cash payment, but can sometimes be sold in exchange for an ongoing royalty and may even include a license-back clause that allows the previous owner to use the patented technology.

The right patent or patent portfolio can be extremely valuable and fetch high prices when sold. For example, a 2012 study of public patent sales found that the average price per patent was $374,000.[1] Individual patents can sell for millions of dollars if they cover important innovations in a large market. Companies with robust research and development programs and large patent budgets are more likely to generate these top assets, but small inventors can still create patents that sell for record prices. For example, in 1997, Juliette Harrington, a mother of three living in a small New Zealand coastal town, filed for a patent that broadly covered online shopping carts for Internet shopping. Although such virtual shopping carts are now universal to online vendors, when the Internet was in its infancy, such a function had not yet been contemplated. After issuing in 1999, the patent subsequently sold for a record $4 million at auction.

Having a patent portfolio with many assets in a specific technology area can be more valuable as a group than any one patent would be individually. Buying such targeted patent portfolios can clear the way for expansion into new markets or can further solidify the exclusivity a company may already have in a given market. Selling under-used patent assets can be a great source of capital. For example, in 2012, after years of decline due to the disappearing photographic film market, the struggling Eastman Kodak company sold 1,100 of its digital-imaging patents for $527 million to raise capital for its restructuring efforts.

BOOK VALUE OF PORTFOLIO

The inherent value of a patent portfolio that could be realized in a sale can be leveraged in other ways, including as collateral for loans. Just

---

1. Calendar Year 2012 Patent Value Quotient, IP Offerings, *available at* www.IPOfferings.com

as banks will lend money based on equity in a home, banks will also lend money based on the value of patents and pending patent applications. Like loans that use a house or vehicle as collateral, patent-based loans include a security interest in the collateralized patents that allows the bank to take and sell the patents if the borrower defaults on the loan. Although this creates a risk of losing patents, the interest rates for such secured loans are substantially lower than unsecured loans. In addition to, or as an alternative to selling patents, patent loans are an excellent source of capital for companies in need.

For example, in addition to selling a group of patents to raise money, Kodak also used another 7,800 of its patents as collateral for a $950 million loan from Citicorp in 2012. This loan and the sale discussed above, provided Kodak with nearly $1.5 billion, which was largely credited for the company's successful emergence from bankruptcy in late 2013.

In addition to serving as a source of collateral for loans, patents can represent a significant amount of value in a company. In fact, for start-ups that are yet to realize positive cash flow, patents often define nearly all of the value in a company. An increased valuation can help attract investors and justify a higher price for company stock. For startups, a rule of thumb used by some investors is that each patent or patent application increases a company's value by $1 million. Accordingly, by making smart investments in a patent portfolio, companies can significantly boost their valuations.

Given this more-is-more perception of patents, I am surprised by how many startups fail to make adequate investments in a patent portfolio, especially given that the cost of several patent applications need not be substantially greater than the cost of just one if planned correctly. The issue is that few entrepreneurs realize that a single product can easily be the source of several patents, so the possibility of multiple patents is never considered.

POSITIVE MARKETING BENEFITS

Along with providing an exclusionary right that directly translates into value through a monopoly, licensing, a sale, or as loan collateral, patents are also an extremely powerful marketing tool. Patents make products more attractive to customers and make companies more attractive to investors and other business partners.

First, having patents is a signal to the public that a company is substantially more innovative than those without patents. A common perception is that only the most elite of inventions will be eligible for patent protection and that patents are issued only after extensive testing by top government engineers or scientists. The more patents a company has is typically seen as being directly proportional to its capacity to develop exceptional innovations.

By carefully publicizing the important steps of developing a patent portfolio as part of an ongoing marketing strategy, companies can project an image of being leaders in a given technology area. Such a positive image can be attractive to potential customers and can influence a decision to choose one product over another. Consumers prefer to buy products that they perceive as having special qualities and added value, and are often willing to pay a premium for such products over similar competitors.

For example, you may be familiar with marketing statements that tout a product for being "awarded" one or more U.S. patent, and graphics that show patent numbers along with the official ribbon copies of these patents. Marketing like this is evocative of winning a blue ribbon in a contest, where awards are limited to only the top competitors. Just as customers would prefer to eat at an award-winning restaurant or use a top-rated service provider, they also prefer to buy products that are recognized as being innovative and at the top of the market. Potential business partners and investors also see patents as indicative of a top company with best-in-show products.

However, as discussed in detail in the coming chapters, this is far from the reality of how patents are examined and issued. Any patent application that sufficiently describes an invention that is new and nonobvious compared to existing technology will be patentable. There is no requirement that the invention be important, special, or even commercially viable. Even silly and simple ideas can be patentable.

For example, in 2002, five-year-old Steven Olson of St. Paul, Minnesota, was granted a patent on his invention for a "method of swinging on a swing." Steven's father, Peter Olson, was a patent attorney and promised his son that if he could come up with a new invention, Steven could file for a patent on it. Steven came up with a new method of swinging on a swing in which a user positioned on a standard swing suspended by two chains from a horizontal tree branch induces side to side motion by pulling alternately on one chain and then the other. The father-son team filed their patent application in 2000 and, like most patent applications, it was initially rejected by the patent examiner. But after a bit of arguing, the examiner was ultimately convinced that the invention was new and non-obvious in view of existing swinging technology, and patent 6,368,227 was allowed to issue.

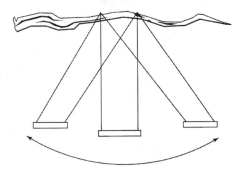

FIGURE 2.2   Figure 2 of U.S. Patent 6,368,277 showing a method of swinging on a swing.

This illustrates that, contrary to popular belief, inventors can patent even the simplest of inventions if they are new and non-obvious. While many patents do indeed protect ground-breaking inventions, one should keep in mind that this is not always the case.[1] However, when it comes to legitimate products and businesses, the public almost universally views patents as being a hallmark of important and valuable inventions. In coming chapters, you will learn how to cost-effectively build a patent portfolio that will attract customers, investors, and busi-

---

1. More examples of weird, silly and crazy patents can be found at www.PatentsDemystified .com.

ness partners alike, regardless of how many major innovations you have actually developed. On the other hand, you will also learn how to read and evaluate patents so that you know whether they protect significant inventions or are simply marketing tools.

PATENT PUBLICATIONS

When a pending patent application publishes, an official copy of the patent application becomes available and searchable on the USPTO website (www.uspto.gov). It can be found by searching for various identifying data including inventor name(s), company name, application number, title, and keywords that may be present in the patent application text. Any documents filed by the applicant or patent examiner during the examination process also become publicly accessible when the application publishes and any new examination documents are immediately accessible from then on. If the patent application is allowed by the examiner and issues, a final version of the issued patent along with its patent number becomes publicly available.

INSTILL FEAR IN COMPETITORS

While seeing patent pending or a patent number on a product creates goodwill in the eyes of many, it evokes fear in competitors and companies looking to enter a market. The patent pending designation indicates that a patent application has been filed for a given invention, but has not yet issued into an enforceable patent. The ambiguity in this status is extremely valuable. By default, patent applications are held in secret for at least 18 months from when they are filed, so there is no way to lookup the patent application associated with a given product during that period. Once patent applications do become public, they are still difficult to access. Without a document to review, "patent pending" could mean anything and the true scope of protection being sought is unknown and undeterminable. The patent could be directed to a small portion of a given product or could be directed to the product generally. The patent could give strong protection or could be completely worthless. Even if a published patent application can be found that corresponds to the patent pending designation of a given product, reading and understanding what the patent application is directed to is extremely difficult without the assistance of a patent attorney. Moreover, the protection afforded often changes during the examination before the patent issues, and in many cases a patent will never issue at all and no enforceable rights will ever be granted.

The opaque nature of the patent process makes it exceedingly difficult for competitors to make an educated decision about whether it is safe to copy or otherwise use technology that is marked as "patent pending." Even sophisticated companies with expert patent counsel have a difficult time making such assessments and it is nearly impossible for companies working on their own. The ambiguity of patent pending therefore creates a strong deterrent for existing competitors and companies considering whether to enter a given market. Instead of the risk of having an operation or product line completely shut down and being liable for damages, many parties simply choose to avoid using patent pending technology if possible.

Once a patent issues and the scope of patent protection is solidified, understanding what is actually protected by a patent remains difficult and requires costly interpretation by a patent attorney. Designing around a patent can therefore be prohibitively expensive for many companies and, if a reasonable license is not available, the best choice is to remain out of the market completely or to tiptoe widely around the perceived bounds of the existing patent landscape. The smart business choice is to, instead, focus on products or services that do not come with the risk of patent lawsuits.

Both patents and pending patent applications therefore provide significant exclusivity in a market and over specific technology simply by scaring away competitors. While some of these parties may have intended to produce a product that would have actually infringed a patent, others will be scared away from making similar products—even non-infringing ones—simply because they are unable to understand what is truly protected by a patent. In this way, patents give protection that can be much broader than the actual granted rights.

DEFENSIVE PUBLICATIONS TO BLOCK COMPETITOR PATENT APPLICATIONS

At some point, patent applications publish and become available to the public. This occurs by default 18 months after the earliest filing date and also when a patent application issues as a patent. In addition to providing effective notice to would-be infringers that you have patent pending or patented technology, such publications can also be used against your competitors who are trying to patent similar technology. As discussed in more detail in the coming chapters, during the examination of a patent, the patent examiner will search for technology that is already known and make a determination of whether the claimed invention of the application is new and non-obvious. Examiners can theoretically use any type of technology disclosure to make this determination, including actual products, scientific publications, news articles, blog posts, or even a Kickstarter campaign. However, because of the extremely limited amount of time that patent examiners have to search for such technology disclosures, in the vast majority of patent examinations, the examiner will only search through patent publications of issued or pending applications. Unfortunately, this means that patents are regularly allowed that are over-broad or completely invalid because the examiner was unaware of highly relevant information that would have otherwise provided a basis for rejecting the patent application.

Although it is possible to invalidate a patent by challenging it in court or through USPTO proceedings, the process is extremely expensive and issued patents are given a presumption of validity, which can be difficult to overcome. As a result, even patents that are clearly invalid can be successfully asserted against you and cause serious problems, especially when the aggressor is well capitalized. In fact, I have seen the consequences of this on a regular basis when defending clients against patent lawsuits. Despite overwhelmingly clear cases for invalidity and non-infringement of a patent asserted against them, clients often still choose to settle the case instead of paying the cost of pursuing a lawsuit. Unfortunately, it is not uncommon for such settlements to be in the hundreds of thousands of dollars.

For example, consider the case of one defendant I worked with who had been selling a line of products for a number of years before a plaintiff competitor filed for and subsequently received patents that covered the defendant's product line. Had a patent examiner been aware of the defendant's earlier line of products when examining the plaintiff's later patent applications, the patents would never had been allowed. As is common in most patent examinations, the examiner did not have time to look for products in the marketplace and instead focused on published patent applications and issued patents. When the plaintiff sued the defendant for patent infringement based on a new line of products that the defendant was selling, the defendant had an outstanding case for invalidating the asserted patents given the earlier line of products that the examiner had overlooked. Unfortunately, the cost of invalidating the patent in court or at the USPTO was too much, and the defendant decided to settle the case for a high six-figure price, despite being right in the eyes of the law.

This whole situation would have been avoided had the defendant originally filed for patent protection on its earlier line of products. The patent examiner would have easily found the patent application publications associated with these products and would have rejected the later applications by the competitor. Filing for patents on technology therefore provides important ammunition that can be used by patent examiners to reject, or at least severely restrict competitor patents in the same field, and often to the same technology.

PATENT ARMS RACE

One of the best ways to defend against the patent arsenals of your competitors is to have a patent arsenal of your own. When considering whether to file a patent lawsuit against a potential infringer, patent holders are less inclined to move forward when an infringement countersuit is a distinct possibility. In situations where a pair of competitors share an intellectual property landscape peppered with respective patents, the best business decision is often to peacefully coexist instead of sparking an expensive patent war that only depletes the capital resources of both sides. Such companies often choose to cross-license instead of litigate. In other words, these competitors will give each other licenses to the patents at issue in their respective portfolios.

The importance of having a patent portfolio to fend off aggressive competitors is well illustrated by Google's acquisition and then subsequent sale of Motorola Mobility. In its first major expansion into a market aside from Internet search and services, Google unveiled its Android operating system for smart-phones in 2007. Almost immediately, Google started being sued for patent infringement by the dominant players in the market, such as Apple and Microsoft, who had sizable patent portfolios on numerous aspects of smart-phone hardware and software. Despite being a large and highly innovative company, Google had a patent portfolio of only about 1,000 patents at the time—a tiny portfolio compared to similar companies of the same size. In fact, Google executives were openly hostile to obtaining patents and specifically chose to make minimal investments in a system in which they did not believe. This strategy managed to work against competitors in the Internet space, but Google was exposed and outgunned as it entered the litigious mobile handset market, where a robust patent portfolio was a necessity. After being beat-up by patent lawsuits for several years, Google bought Motorola Mobility for $12.5 billion in 2011, with the primary reason for the acquisition being the 17,000 mobile-related patents that Motorola Mobility owned. This new patent portfolio was used to fight back against competitors, and helped propel Android to become the most popular operating system as of mid-2012. While retaining the patents it gained in the original acquisition, Google sold Motorola Mobility to Lenovo for $2.9 billion in early 2014. After considering the cash that came and went with these deals, Google spent $7 billion, or approximately $410,000 per patent, for an important patent arsenal that allowed it to emerge as the dominant player in the smart phone market.

PATENT VALUE VS. ACQUISITION COSTS

Aside from the granted right to exclude others from making, using, selling, or offering to sell claimed technology, patents provide numerous other benefits that often go unappreciated by inventors and companies seeking patent protection for their ideas. Unfortunately, this often results in significant under-investment in patent assets because a true value is hard to calculate. For example, as with many marketing efforts, it can be difficult to accurately assign a value to the benefit gained by advertising patent assets. The number of sales of your product over a competitor that can be attributed to patent marketing is hard to quantify—patents are typically only one of many factors in consumer choice. Also, the number of companies that never entered the market or entered on a limited basis due to fear of patents is nearly impossible to quantify. Even if you could determine actual instances of this occurring, putting a dollar value on it is difficult. Similarly, you may know that patent assets were one factor that attracted certain investors, business partners, or executives to work with your company, but again, placing an exact dollar value on this not an easy task. Regardless, it is clear that there is significant value gained through patent protection, and almost certainly a substantial value over the cost of obtaining patents.

The potential value of patent assets in a sale, license, or as collateral for a loan can also be difficult to assess given the unique nature of patents and the markets that they are relevant to, but it is not uncommon for patents to be worth tens if not hundreds of times what it costs to acquire them. In the worst case scenario, it is not unreasonable to expect to recover the costs of getting a patent by selling it. In the coming chapters, I'll show you how to cost-effectively navigate the patent system and maximize the potential of growing patent assets that could be worth millions to your company.

CHAPTER 2 SUMMARY

- An issued U.S. patent provides, for a limited term, the right to exclude others from making, using, offering for sale, or selling the claimed invention throughout the United States or importing the invention into the United States.
- Patent owners can maintain exclusivity in the marketplace for themselves or license the technology to others in exchange for cash or other benefits.
- Patents and patent applications can be sold for a profit or used as collateral in business loans.
- Patents and patent applications have a positive marketing benefit for companies and also create a deterrent for competitors.
- Publications of issued patents and patent applications can prevent others from obtaining patents on similar and related technology.

Overview of the Patent Process

Before discussing the pieces of the patent process in-depth, it helps to get an idea of how the patent process looks as a whole. This chapter introduces a skeleton of a general timeline and steps in getting a patent on which the following chapters build and expand in generally the same order.

INVENTION AND DEVELOPMENT

Once the spark of invention occurs, a clock begins to tick. The United States now gives priority to the first inventor to file a patent application and patent rights can be lost completely if an inventor is not careful about making public disclosures, public uses, and offers for sale of an invention. Therefore, a patent application should be filed as soon as possible after an invention is conceived so that patent rights are not lost to others who filed first or as a consequence of releasing a secret invention too soon. At the same time, filing a patent application too early can be detrimental—sometimes an invention needs to be further developed before it is ready to be patented. In some cases, a business simply needs time to develop a product, raise capital, or explore the viability of a business or product before patent applications can be filed.

The timing of filing a first patent application is dependent on the specific invention, the type of technology, and the business plan of the applicants or company. In some cases, patent applications are filed when an invention is little more than broad strokes of an idea sketched out in the mind of an inventor. In other cases, an invention may be in development for years with numerous generations of prototypes being built, rejected, and revised before a first patent application is ever filed. Knowing when filing for a patent is too early or too late is often counter-intuitive, so a trusted patent attorney should be consulted early in the process to prevent an inadvertent and unnecessary loss of patent rights.

CHOOSE AND MEET WITH A PATENT ATTORNEY

A patent attorney should be as carefully selected as any business partner or advisor would be. Patent attorneys do more than just draft and file patent applications—they should be an integral and active advisor in planning and executing a personalized intellectual property strategy for your business. The right patent attorney can add millions of dollars in value to a company, whereas another, less qualified attorney can create a drag on a company that ends up killing it. Finding a trusted patent attorney that understands your business and invention and has experience crafting and executing successful patent portfolios is therefore essential.

Once a patent attorney has been selected, the inventors or business promoters work with the attorney to map out an initial plan, which includes a complete disclosure of the invention (or inventions) for which protection is being sought. The patent attorney will also make an assessment of whether the invention is ready for patenting; whether prior art searching should be done; and what type of patent application should be filed to start the patent process.

As discussed in detail in the coming chapters, the patent process can begin with filing a provisional or nonprovisional utility patent application, and choosing which one to begin with depends on the specific needs of the business and the type of invention technology. The nonprovisional application is the application that is substantively examined by the USPTO and will ultimately mature into an issued patent if it is deemed patentable. In contrast, a provisional application is merely a placeholder for a subsequent nonprovisional application. Provisional applications are *not* substantively examined by the USPTO and automatically expire one year from filing. A nonprovisional application must therefore be filed within this one-year term that claims benefit of the provisional application. Accordingly, there are two primary options for the utility patent process: (1) file a provisional patent application first and then file a nonprovisional patent application within one year or (2) skip the provisional application step and file a nonprovisional application first.

If a provisional application is filed first, the application is drafted ideally over the course of a few weeks and then filed with the USPTO. Since this provisional application expires one year from its filing date, a nonprovisional application must be filed no later than this one-year deadline. Material of the provisional application can often be substantially used when drafting the nonprovisional application, but revisions and additional material is typically required. On the other hand, the patent process can also begin by first filing the nonprovisional patent application and the step of filing a provisional application and waiting up to a year is simply skipped. Regardless of which type of patent application is used to start the utility patent process, the invention is considered "patent pending" when the first application is filed.

EXAMINATION PROCESS AT THE USPTO

The nonprovisional utility application is the application that will hopefully mature into an issued patent. This application begins its life at the

USPTO with some preprocessing to assure that it meets basic filing requirements, and is then placed in line to be examined by a patent examiner. Because of the massive backlog at the USPTO, an examiner will not have the opportunity to begin examination of the application for approximately one to three years from its filing date.

Applicants become aware that the examination process has begun when the examiner mails an "Office Action" to the patent attorney that sets forth various types of application rejections. These rejections are formed by the examiner after a prior art search and a check to see if the application complies with a host of patent laws and regulations. As discussed in detail in the following chapters, such rejections are a natural and anticipated part of the examination process. In fact, it is extremely rare to receive an immediate allowance of a patent application, and it is actually cause for concern if an application is allowed in the first Office Action. The examination process should be viewed as a negotiation, with the patent attorney being a negotiator who carefully sets up an initial offer in the filed application and then negotiates the best possible patent rights during examination.

After analysis of the Office Action, the patent attorney prepares and files a written response within the required deadline, which makes amendments to the application and/or makes arguments against the rejections asserted by the examiner. The examiner then responds with another Office Action. This back and forth in the form of Office Actions and attorney responses continues until the examiner allows the application or the applicant gives up and abandons the application. If the examiner allows the application, however, the applicants pay final fees and the application then issues as an enforceable patent. The timeframe of responding to an Office Action is usually within three months of its mailing date and it can be several months or even years before the examiner considers the response and continues examination. Therefore, depending on the number of rounds of Office Actions and responses, the time from the beginning of examination to issuance of a utility patent can often be one to four years or more.

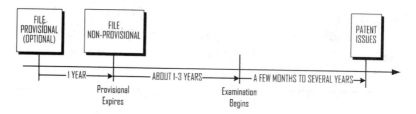

FIGURE 3.1   Example timeline of obtaining a
United States utility patent.

Before a patent application issues as a patent, the applicant has the opportunity to file one or more additional applications related to the first patent application. These additional applications—called continuation or divisional patent applications—can be extremely useful in protecting multiple inventions that are described in a single patent application. Having multiple patents that protect different parts of a product, or protect it in different ways, is one strategy for growing value of a patent portfolio, and can be strategically beneficial in patent litigation. Such advanced patent strategies are discussed in more detail in coming chapters.

PRE- AND POST-ISSUANCE ACTIVITIES

Patent rights can only be asserted in court against infringers once the patent issues, so issuance of the patent opens the door for enforcement against competitors and other infringers. Patent lawsuits can stop infringers from making, using, selling, or offering to sell the patented invention within the United States. However, once the first patent application is filed an applicant should feel free to begin marketing, selling, and otherwise publicly exploiting the patent pending invention.

FOREIGN PATENTS

U.S. patents only protect an invention within the bounds of the United States, so, if protection is desired in other countries, applicants must file for patents in each foreign jurisdiction separately. Regional patent offices exist for the European Union and certain groups of countries in Africa and Eurasia, but such regional patents must still be validated in each country where patent protection is sought. Foreign patent filings must be made within a year of the first U.S. application (either provisional or nonprovisional application), otherwise foreign patent rights are generally forfeited.

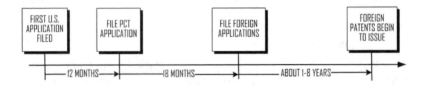

FIGURE 3.2   Example timeline of the PCT foreign patent
application process

One option is to file directly in each desired country by the one-year deadline discussed above. Alternatively, applicants can file what is known as a Patent Cooperation Treaty (PCT)[1] application, which acts as a placeholder for future foreign filing in PCT member countries.[2] Filing a PCT application has the benefit of providing at least 18 additional months before applicants need to choose foreign countries in which to file patent applications. Plus, the PCT option can make the foreign patent process less expensive overall if several foreign patent applications are ultimately filed based on the PCT application.

The process of obtaining foreign patents is extremely expensive and is therefore a poor investment for the vast majority of companies. The foreign patent process includes having a United States patent attorney coordinate with foreign patent attorneys in each country or region, which can cost double what the process costs in the United States. Additionally, non-English-speaking countries require a complete

---

1. The Patent Cooperation Treaty of 1970 is an international treaty that provides a unified procedure for filing patent applications in each country that has joined the treaty.
2. Nearly all major foreign jurisdictions are PCT member countries. Notable exceptions include Taiwan and Argentina.

translation of an often lengthy patent application, and the technical subject matter of patent documents makes these translations costly. Moreover, unlike the United States, most foreign jurisdictions require increasing yearly maintenance fees to keep pending patent applications alive and to keep issued patents in force.

Aside from large global companies and businesses that already have a substantial market established in specific foreign markets, companies should focus their patent budget on growing a U.S. patent portfolio. For this reason, foreign patents are discussed a bit in coming chapters, but the primary focus of this book is on the U.S. patent process.

CHAPTER 3 SUMMARY

- Applicants can begin the patent process by filing a provisional or nonprovisional application and both provide patent pending status.
- Provisional patent applications are placeholders that *are not* substantively examined and automatically expire one year from filing.
- A nonprovisional application must be filed within this one-year window to preserve the earlier priority date of the provisional application.
- Nonprovisional patent applications *are* substantively examined at the USPTO and will hopefully mature into an issued patent.
- Examination of a nonprovisional patent application typically begins one to three years after filing.
- The examination process begins with the assigned patent examiner conducting a prior art search, which leads to a negotiation between the patent attorney and examiner that ideally results in the application being issued as a patent.
- It is possible to also file patent applications in foreign countries, but this is not recommended for most businesses.

4

Prefiling Procedures and Considerations

"A person shall be entitled to a patent unless . . . the claimed invention was patented, described in a printed publication, or in public use, on sale, or otherwise available to the public before the effective filing date of the claimed invention."

—Title 35 of the United States Code, Section 102(a)

After coming up with a new product or business idea, many inventors get stuck. They recognize they have something to share with the world but, at the same time, they are afraid that if they share their idea with others it might get stolen. In other words, they want to be able to develop, sell, and profit from their inventions while being assured that others will not be able to copy their products without permission and without acknowledgement of their inventorship. While it is impossible to absolutely guarantee security of an idea or product, there are several important steps that inventors should take to avoid forfeiting their rights or creating an opportunity for others to misappropriate an invention before it is ready for patenting.

THE NEED TO KEEP AN INVENTION SECRET

The first rule of inventing is that inventions should ideally be kept completely secret until a patent application is filed. In a perfect situation, a patent application would be filed immediately after the invention without disclosure to anyone outside of the inventor or group of inventors. However, the reality is that absolute secrecy is not always practical or possible. Most inventors want to develop the product, raise money, consult experts, and do some market research before they invest time and money into filing patent applications. Accordingly, being careful of how disclosures are made is imperative. Certain activities can result in the irreparable loss of patent rights, while other activities can be reasonably safe if done correctly.

POTENTIAL LOSS OF PATENT RIGHTS (U.S. AND FOREIGN)

Even after some changes to the patent laws in 2013, inventors must generally file for a U.S. patent within one year of the first public disclosure, public use, or offer-for-sale of an invention. Otherwise, the invention is effectively dedicated to the public domain and anyone is free to use it. On the other hand, in the vast majority of foreign jurisdictions, all patent rights are lost *immediately* upon a public disclosure, public use, or offer-for-sale of an invention—even if such activity occurs in the United States. As discussed in more detail in the coming chapters, foreign patent protection is not recommended for the majority of companies, but this does not mean that foreign protection should be forfeited in favor of the one-year grace period afforded by the United States.

Foreign protection may be important to potential investors or business partners in the future, and having the option for foreign protection may make the difference between closing a deal or not. For example, while a small startup may not initially have sufficient capital to support foreign patent filings, a venture capital firm or acquiring company may be able to provide sufficient capital and may also have market penetration in foreign jurisdictions that would justify building a foreign patent portfolio. Prematurely giving up the option for foreign patent protection is therefore an unwise decision for even the smallest company, especially given that limiting public disclosures, uses, and offers-for-sale until filing for a United States patent is often a relatively small sacrifice to make.

Additionally, although the changes to United States patent laws in 2013 maintain a one-year grace period for many activities, these new laws are yet to be completely interpreted by the courts. Therefore, it is not entirely settled as to what activities will start the clock ticking on the one-year grace period; what activities are safe; and what activities will immediately forfeit patent rights. To be safe, inventors should therefore avoid certain public disclosures, uses, or offers-for-sale, even if they might only trigger the one-year clock in the United States. Accordingly, the following is a discussion of activities that should be avoided until a patent application is filed.

PUBLIC DISCLOSURES

A disclosure can be considered "public" even if it is to a single person or the posting of a written description of the invention where it *might* be viewed by a member of the public. In other words, "public" does not require the information being disseminated to a large number of people and, in some cases, proof that even a single person received the information may not be required.

For example, in 1977, Peter Foldi, a graduate student at Freiburg University in Germany was granted a PhD in chemistry after presenting his doctoral dissertation to the department of chemistry and pharmacy. A single hard copy of his original dissertation was made available in an isolated special dissertation section of the university library. Along with the multitude of dissertations that had come before him, Dr. Foldi's work was indexed in the dissertation section of the library's user catalog. Although there was no evidence that a single person had ever used this catalog to search out Dr. Foldi's contribution to the science of enzyme chemistry, the simple fact that the dissertation *could* have been accessed by the public was deemed to be a public disclosure that later served to invalidate a patent on a similar enzyme.[1]

---

1. See *In re* Hall, 781 F.2d 897, 228 (Fed. Cir. 1986).

In view of cases like this, information about inventions should not be published anywhere that the general public could theoretically get access to it. This includes public webpages, blog posts, videos, images, or audio recordings—even if there are no active links to such content or if the public is unlikely to stumble upon the content.

Oral disclosures can qualify as public disclosures just as much as publications can, so discussing inventions with outsiders should be avoided if possible. Where oral disclosures are unavoidable or necessary, however, it may be possible to do so without unduly compromising patent rights. In many cases, if confidentiality can be established for a given disclosure, it will not be considered a public disclosure. Establishing confidentiality, discussed in detail later in this chapter, can be used to protect certain public uses of an invention that would otherwise harm patent rights.

PUBLIC USE

In addition to publications and oral disclosures, using an invention in public can also harm patent rights. Again, in many cases, a use can still be considered "public" if it involves a single person—even without proof that a single person ever saw the invention being used. Moreover, using an invention in public can be harmful even if the invention is hidden from view or the function of the invention is not readily discernable.

For example, in 1873, Samuel H. Barnes was issued a patent on improved springs for ladies corsets that provided for improved strength, flexibility, and elasticity. These springs, or "steels" as they were known, were sewn into slots within the fabric of a corset and clothing was worn over the corset. The springs were therefore not outwardly visible while being worn. Mr. Barnes gave his wife and at least one of her friends a set of his new corsets springs, which they proceeded to wear in public for well over a year before he filed his patent application. These innovative corset steels worked so well that his wife and her friends would tear them from corsets that had worn out, and sew them into new corsets. Unfortunately, when Mr. Barnes tried to assert his patent against an infringer, it was invalidated because wearing the steels in public was considered a public use.[1]

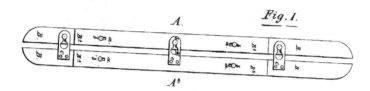

**FIGURE 4.1   Corset-Springs Illustrated in S. H. Barnes'
Patent RE5216**

Cases such as this demonstrate that inventions should not be used in public until a patent application has been filed. As this case illustrates, even a hidden use that occurs under clothing, or is otherwise not viewable, can still harm patent rights if the use occurs in a public place.

This need for complete secrecy applies even more so to patentable methods such as a process of making a product or a patentable machine that makes a product. Patent rights to an innovative method or machine can be lost if the result or product generated by it is made public. For example, suppose a new process of making chocolate bars is developed in secret along with a set of machines that implement the new method. Both the method and machines could be patentable. However, if the chocolate bars produced by this secret facility are made public or sold publicly, then the ability to protect the innovative method and machinery may be lost.

Applied to software inventions, suppose that a public website runs a novel search engine algorithm that provides for faster and more relevant search results. Even if the code that implements this search engine is not publicly accessible or if it is impossible to reverse-engineer the

1. *See* Egbert v. Lippmann, 104 U.S. 333 (1881).

novel algorithm, the fact that the website is available to the public would still harm patent rights to this invention. Exploiting an invention in any public way is therefore inadvisable and should be avoided until a patent application is filed. This includes public disclosures, public uses, and offers-for-sale.

OFFERS-FOR-SALE

In addition to compromising patent rights by using an invention in public or disclosing the technology publicly, patent rights can be harmed by offering a patentable product for sale or by offering to sell a product produced by a patentable machine or manufacturing process. To be clear, there is no requirement that an actual sale of such a product occurs—merely offering it for sale is sufficient. Also, unlike the detrimental disclosures or uses discussed above, such an offer-for-sale need not be public. A confidential offer to sell a product can be sufficient to harm patent rights.

A conventional advertisement for a product or actual sale of a product to the public in exchange for money, goods, or services is a clear example of an activity that can be detrimental to patent rights. However, other types of sales and business relationships can be equally harmful—even where a sale is from a manufacturer to the patent owner or from the patent owner to distributors. For example, patents were invalidated in one case because the innovating company entered into a requirements contract with a manufacturer to provide one million units per year of a product that was the subject of a patent.[1] A sale between two distinct companies, even closely related companies, can still be problematic.

For example, before patenting its novel K-2000 surgical blade, the Brasseler Company entered into an agreement with DS Manufacturing to make the blades and sell them to Brasseler so that Brasseler could then package and sterilize the blades for resale. A sale of more than 3,200 blades from DS Manufacturing to Brasseler was sufficient to invalidate a patent that Brasseler had received, even though

---

1. Special Devices Inc. v. OEA Inc., 270 F.3d 1353 (2001).

the blade technology was jointly developed by inventors from both companies.[1]

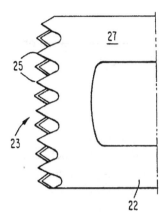

FIGURE 4.2   Drawing from the Brasseler Company's surgical blade patent.

Given that even small disclosures, uses, and offers-for-sale can potentially forfeit patent rights, it is imperative to work closely with a patent attorney when planning the development and exploitation of a new invention. Market testing, finding operating capital, cultivating business partnerships, developing a prototype, and mass production of a product are all necessary parts of most businesses, but such activities nonetheless risk loss of patent rights if not done correctly. By engaging a patent attorney early in the process, a plan for maintaining patent rights can be formulated that correctly balances the need to protect patent rights with the goal of building a business.

However, if an invention has already been disclosed, marketed, or publicly used, inventors should not be disheartened or assume that all rights are already lost. Instead, a patent attorney should be consulted as soon as possible to get an assessment of what rights are still available and what steps can be taken to preserve any patent rights that still

---

1. Brasseler v. Stryker Sales Corp., 182 F.3d 888 (Fed. Cir. 2001).

remain. In many cases, an exception may apply or an invention may still be within the one-year grace period.

SECRECY

By maintaining secrecy around disclosures or uses of an invention, it may be possible to avoid the negative consequences of "public" disclosures or uses as discussed above. However, this secrecy exception does not apply to sales or offers-for-sale. Confidential sales or offers-for-sale are just as detrimental as public ones.

Ideally, secrecy should be formalized with a contract. Known as a nondisclosure agreement or NDA, such contracts create an obligation for recipients of information to keep that information secret. An NDA may even impose penalties on parties that breach the contract if they make secret information public. At the same time, NDAs may be redundant or impractical in certain situations and secrecy is best maintained informally.

NONDISCLOSURE AGREEMENT (NDA)

A formal NDA is the next best thing compared to not making invention disclosures at all. Contrary to popular belief, there is no such thing as a "standard" NDA, but many NDAs will have similar clauses and terms. For example, NDAs typically define who the parties are; a date or timeframe when confidential information will be shared; a general description of what types of confidential material will be shared; a term for which the secrecy must be maintained; and consequences for the receiving party if secrecy is not maintained. An NDA can be a stand-alone contract or such terms may be part of other contracts, such as an employment agreement, contractors' agreement, development agreement, or manufacturing agreement.

The main benefit of an NDA comes from its formality. For most signers, it is a clear indication that the party conveying confidential information considers it to be extremely important and that unauthorized disclosure of the information being imparted to them would result in serious consequences. The act of pausing to read and sign an NDA makes it less likely that the receiving party will forget the requirement of secrecy sometime in the future.

On the other hand, although an NDA is a best choice from a legal perspective, NDAs are often a poor choice from a business perspective. For example, consider that the ultimate purpose of NDAs is to facilitate collaboration with others, including important potential business partners, investors, distributors, manufacturers, and the like. Unfortunately, NDAs can actually stifle these relationships or even block them completely, which can be substantially more detrimental to a growing business compared to the theoretical risk of employing more informal methods of maintaining secrecy. Networking is one of the most important aspects of an emerging business, yet far too many startups wither and die because they are overly cautious about sharing their ideas with others or alienate important connections by negligently using NDAs.

For example, many investors such as venture capitalists or angel investors refuse to sign NDAs and will not even meet with companies that require them. Moreover, asking these parties to sign an NDA is often seen as an indication of an inexperienced business person and may ruin a potential financing deal before the meeting has even begun. This unwillingness to sign NDAs is not because investor groups want the option of stealing inventions that are pitched to them. Instead, it is because these sophisticated investors know that NDAs are extremely difficult to enforce in practice, even if a breach has occurred, and that NDAs also impose unacceptable liability in many cases. The reality is that, while NDAs help to prevent forfeiting patent rights, they often only provide an illusory sense of security against other parties stealing the disclosed idea. The best way to prevent others from stealing your idea is by first filing a patent application, not by having them sign an NDA. This is one of the many reasons filing a patent application as soon as possible should be a top priority.

Despite their potential downside, NDAs are still an important tool for any business that wants to keep inventions secret before filing a patent application. However, just as any carpenter or plumber knows, the right tool must be selected based on the job at hand. NDAs should not be completely avoided just as much as they should not be used by default. NDAs should only be used in the right circumstances, which require good business judgment and knowledge of customary dealings for certain types of relationships and even of customary dealings for specific companies or individuals. In some cases, the other party may be surprised if you *do not* have an NDA for them to sign, whereas in other cases it can be seen as an insult. Trusted attorneys and people experienced with industry practices or specific parties in a field can be good resources if you are unsure of how to tactfully protect yourself when making invention disclosures. Luckily, in many cases, there are suitable alternatives to NDAs for protecting your patent rights before filing a patent application.

INFORMAL, INHERENT, OR IMPLIED CONFIDENTIALITY

As discussed above, NDAs formally establish a requirement for secrecy with a written contract. However, when it comes to preventing "public" disclosures that will harm patent rights, such formality is not always necessary. For example, where secrecy is implied, inherent, or informally established between parties, this can be enough to prevent a given disclosure from being considered a "public" disclosure that can result in forfeiture of patent rights.

Informal secrecy can be established in many ways, including by simply stating that a given disclosure or demonstration is confidential and should not be shared with anyone else without express permission. There are no special or magic words that need to be uttered—the idea is that the receiving party should understand that the information being shared should remain secret.

Inherent secrecy may be present where there is an existing duty of secrecy based on the relationship of the parties. For example, for attorneys, ethics rules such as the attorney-client privilege define that conversations between attorneys and clients, and typically potential clients, must be kept confidential by the attorney. Accordingly, it would be redundant to ask your patent attorney to sign an NDA before meeting to discuss your invention in an official capacity because secrecy is already inherently required. At the very most, you may want to confirm that attorney-client privilege applies to your meeting.

The context and circumstances of an invention disclosure can also create implied secrecy that qualifies as a nonpublic disclosure that does not harm patent rights. For example, in 1972, Larry Nichols received a patent on his new puzzle cube, and in the early '80s, the patent was asserted against the world-famous Rubik's Cube. (Rubik was able to get a patent on his puzzle cube in 1977, but only in Europe,[1] unfortunately). The defendant argued that Nichols' patent was invalid because several people had seen his prototypes over a year before he filed for a patent. While doing graduate work in organic chemistry, Nichols made several prototypes that were seen by a few close friends, including two roommates and a colleague in the chemistry department. They had each seen the prototype in Nichols' room and were given a demonstration of how it worked.

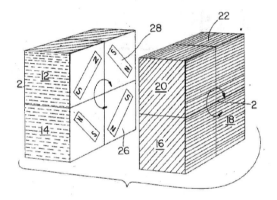

**FIGURE 4.3   Drawing from U.S. Patent 3,655,201
to Larry Nichols.**

Later, after taking a job at Moleculon Research, the company president took interest in the puzzle upon seeing it in Nichols' office. The president received a full explanation of how the puzzle worked and Nichols ultimately sold his rights to Moleculon in exchange for future royalties. In each case where his prototypes were seen and disclosed, Nichols did not get a signed NDA from those viewing the puzzle, nor

---

1. Hungarian patent number HU170062A and Belgium patent number BE887875A.

did he request that they maintain secrecy regarding his invention. However, the court considered these to be secretive disclosures because "the personal relationships and other surrounding circumstances were such that Nichols at all times retained control over [the puzzle's] use as well as over the distribution of information concerning it."[1] The court further found that Nichols "never used the puzzle or permitted it used in a place or at a time when he did not have a legitimate expectation of privacy and of confidentiality."[2]

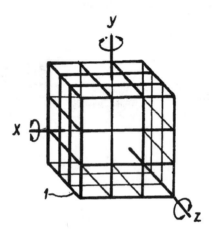

FIGURE 4.4   Drawing from European Patents to E. Rubik.

Although cases like this indicate that invention disclosures may be considered secretive based on the context of the disclosure and relationship of the parties, invention disclosures should still be avoided, and extreme care should be taken if disclosures are absolutely necessary. Any disclosure creates a risk that the informed third parties will disclose the invention to others or share materials provided to them. Regardless of whether such a disclosure is malicious or accidental,

---

1. Moleculon Research Corp. v. CBS, Inc., 793 F. 2d 1261, 1265 (1986).
2. *Id.* at 1265.

patent rights can still be lost as a result of unauthorized third-party disclosures. For example, a patent on a new method of making soap was invalidated because of a public disclosure, even though the public disclosure was made by a third-party that had misappropriated the innovation from the inventor.[1]

Because the law differentiating secretive and public invention disclosures is constantly changing and ultimately decided on a case-by-case basis, the ramifications of any planned disclosure should be discussed and planned with a patent attorney. Regardless, no disclosure should be considered completely safe until a patent application has been filed. Filing a patent application as soon as possible should therefore be a top priority.

WHEN IS AN IDEA READY TO PATENT?

As discussed above, filing a patent application is the only sure way to guard against others stealing an invention or an inadvertent forfeiture of patent rights due to public disclosures, public uses, or offers-for-sale. Filing a patent application once an invention is sufficiently developed is therefore important. The urgency for filing quickly is even more pronounced after recent changes to U.S. patent law.

THE FIRST-TO-FILE SYSTEM

Until 2013, the United States had the only remaining patent system in the world that gave priority to the first party to discover an invention. This so called "first-to-invent" system made it possible for inventors to gain priority if they could prove the earliest conception of a given invention. In other words, a later patent filer that invented first could be entitled to a patent over others that may have filed patent applications earlier. By providing evidence of earlier conception, diligent development of the invention, and timely filing of a patent application, this earlier inventor could win priority over later inventors.

---

1. Lorenz v. Colgate-Palmolive-Peet Co., 167 F.2d 423 (3d. Cir. 1948).

MYTH OF THE MAIL-IT-TO-YOURSELF PATENT

There is a persistent myth still circulating that claims inventors can get a patent by writing a description of an invention and mailing this document to themselves and keeping the envelope sealed. Another variation claims that doing this is equivalent to a provisional patent application or secures other patent rights. To the contrary, self-mailing of invention details only serves as evidence that the writer invented something as of the mailing date. No patent rights are protected by such a mailing and establishing a date of invention is worthless now that the United States gives priority to the first inventor to file a patent application instead of the first to invent. Filing a patent application is the only true way to protect an invention.

However, in 2013 the United States changed its patent system to a "first-to-file" system. Now, instead of giving priority to the first inventor, priority goes to the first inventor to file a patent application. This change makes it even more important to file a patent application as soon as possible. At the same time, a patent application should only be filed once an invention is sufficiently developed and ready for patenting.

TOO EARLY VS. TOO LATE

Many factors contribute to determining when a patent application should be filed and this is ultimately an analysis that should be made in collaboration with a patent attorney. As discussed above, filing too late can result in loss of patent rights because of public disclosures, public uses, or offers-for-sale that originated from the inventor. Additionally, where a technology field progresses quickly and other innovators contemporaneously develop the same or a similar invention and release products, make publications, or file for a patent, such activities can prevent a first inventor from getting a patent. Therefore, waiting to file for a patent increases the risk that patent rights for an invention may be preempted by others or accidentally forfeited by the inventor.

On the other hand, filing too soon can also have detrimental consequences and add unnecessary cost to the process of obtaining patents. As discussed in more detail in coming chapters, filers only get priority to aspects of an invention that are sufficiently described in a patent application. Although filing a patent application at early stages of development protects the invention as soon as possible, further developments will likely require further patent applications so that these changes and additions to the invention will be protected.

In my experience working with startups of various sizes, I find that most innovations change significantly as the inventors refine a product and attempt to bring it to the market. Beginning with an initial theoretical concept and moving to research and development efforts that often include several generations of prototypes, the first idea is often completely alien compared to the finalized prototype that will be the subject of market or user testing. Changes tend to be even more extreme once the product is introduced to the demanding and unforgiving marketplace of consumers.

Although protecting all iterations of an invention during the development process is theoretically the best way to protect intellectual property and build a top-notch patent portfolio, doing so tends to be cost-prohibitive and wasteful for normal startups. Unless a company intends to have a core focus on patents and has a sizable patent budget, filing before the invention or product is relatively solidified is typically unwise.

Filing a patent application too early can also be harmful if changes are subsequently made to the invention and a follow-up patent application is never filed to protect the new aspects of the invention. It is not uncommon for companies to incorrectly assume that a first patent application will always cover their invention regardless of how it develops, and important new innovations are accidentally left unprotected. As a result, a patent that issues may not even cover the product that a company ultimately sells.

Accordingly, forming an early patent strategy with a patent attorney is the best way to strike the correct balance between filing too soon and filing too late. Factors such as the type of invention and unique business plan of a company will come in to play when determining a cost-effective solution to protecting an invention. For example, in some cases, a patent application should be filed immediately after conception, whereas some inventions would benefit from months or years of development before a patent application is filed.

PROTOTYPES

It is a common misconception that having a prototype is required before filing a patent application. In fact, filing a patent application is actually considered equivalent to making a prototype in the eyes of the law. As discussed above, it may be helpful to make a prototype so that a product can be refined and developed before filing a patent application, but for certain inventions and businesses, it actually makes sense to file a patent application before a prototype is ever built. For example, where a prototype is prohibitively expensive to build for the inventors alone and investors are being sought to provide capital for bringing a concept to fruition, filing a patent application at this early stage is necessary. In such cases, the inventors need to be protected while raising capital and having a patent application on file makes the business more attractive to potential investors.

Luckily, the level of detail required for a skilled patent attorney to draft a suitable patent application is surprisingly low for many inventions and, in many cases, the inventors themselves are not even required to have the skills to build a prototype on their own. To provide protection, a patent application must include sufficient detail such that the invention could be made and used by an average person in the technology field. In other words, if the invention can be described to a patent attorney such that he can draft an application that would be enabling to a person of ordinary skill in the art, the invention is capable of being patented. Although the final patent application is often a lengthy and detailed technical document, the amount of information required for a patent attorney to draft this document can be extremely limited. For example, I once drafted a complete and detailed patent application based on nothing more than some crude drawings made on a cocktail napkin. This process of adequately describing an invention to your patent attorney is discussed in more detail in the coming chapters.

ISSUES WITH MULTIPLE INVENTORS

Recent studies have shown that patent applications are increasingly filed by multiple inventors.[1] Although collaboration can be an important part of developing innovations, having more than one inventor of an invention raises some issues that should be addressed as early as possible, and ideally before inventing even begins.

By default, co-inventors each independently own a 100% share of an invention. Each of these inventors has an equal right to the invention, regardless of how much their contribution was, and use of their equal share is not subject to control by the other inventors. For example, suppose three friends come up with an idea for a new smartphone app and eventually get a patent that names them as co-inventors. As the business grows, one of the inventors disagrees with the direction of the company and things become so contentious that he quits to start his own competing company. This new company begins to develop a competing app that is covered by the patent and the new company eventually licenses the technology to Google in exchange for millions in yearly royalties. Unfortunately, the two inventors that were left behind can do nothing to stop their former friend and they are not entitled to a penny of the royalties from his lucrative contract.

Unless these three friends understood and accepted the potential negative consequences of the default ownership structure afforded to co-inventors, then they made a huge mistake by not consciously choosing how their patent rights would be owned. Fortunately, there are easy ways to modify the default ownership created by co-inventorship.

---

1. *See e.g., Inventor Count*, Dennis Crouch, www.Patentlyo.com, July 17, 2013; *Cross-Border Inventors*, Dennis Crouch, www.Patentlyo.com, Nov. 21, 2010; *The Changing Nature Inventing: Collaborative Inventing*, Dennis Crouch, www.Patentlyo.com, July 9, 2009.

ASSIGNING TO A COMPANY

Although there are many possible ways to handle the issue of multiple inventors, the simplest and most common is by collectively assigning patent rights to a company. By doing this, the company then owns the patent application and resulting patent. The rules that define the corporate structure then dictate the use and sale of inventions covered by the patent and also dictate how the patent is licensed, sold, or otherwise exploited. Transferring ownership from inventors to a company makes it possible to consciously choose an ownership and control structure instead of being subject to undesirable default rules.

For sophisticated companies, assignment of all intellectual property rights occurs when an employee signs an employment agreement that includes a provision that rights to inventions are automatically assigned to the company at the moment they are conceived. This standard clause is often present for every employee—from the janitors to the head researchers. Emerging companies and groups of inventors should likewise have similar agreements in place for all parties involved in the company, regardless of whether these employees or contractors plan on making inventive contributions.

WORKING WITH OTHERS

Even after a product has been initially conceived, companies must still be cautious as the product is developed and researched, especially when working with parties outside the company. For example, it is not uncommon for a company to work with outside vendors when making or developing prototypes that embody an invention, and these outside vendors may make contributions to the invention or come up with new ideas that are potentially patentable. As with any inventors, these parties may then have an ownership right in their ideas by default unless they make an explicit agreement that says otherwise. Accordingly, when working with outside parties in developing, researching, or implementing an invention, contractor or collaboration agreements should clearly state how patent rights will be assigned if an inventive contribution is made by this outside party.

TALK TO A PATENT ATTORNEY EARLY

As illustrated throughout this chapter, there are numerous patent issues that arise well before a patent application is ever filed, and even well before an invention is conceived. Although this chapter provides solid guidance for preserving patent rights and establishing a patent strategy, forming a relationship with and consulting a patent attorney as soon as possible can be immensely helpful when planning and growing a business. A trusted patent attorney can provide guidance on avoiding patent pitfalls that are unique to a given business model or technology. Patent attorneys can also assist with formulating a patent game plan that can be followed throughout the invention and development process. Such counsel is cheap and often free insurance against inadvertent loss of patent rights and may even speed up the progression of a company. Finding and engaging a top-notch patent attorney is discussed in depth in Chapter 8.

CHAPTER 4 SUMMARY

- Applicants should plan on filing a patent application before making any public disclosures, public uses, or offers for sale of an invention, which would otherwise cause loss of patent rights.
- Priority is given to the first party to file, so patent applications should ideally be filed as soon as an invention is ready to be patented.
- By default, inventors own the rights to their inventions, so patent assignment agreements should be signed in favor of a company even before inventing activities begin, especially where there are multiple inventors.
- Talk with a patent attorney as early as possible to craft a preliminary patent strategy and to determine when the timing is right to file a patent application.

Patent and Prior Art Searching

"If you know the enemy and know yourself you need not fear the results of a hundred battles."

—Sun Tzu, The Art of War

During examination of a patent application, a patent examiner conducts at least one prior art search to determine whether the claimed invention is new and nonobvious. In certain situations, inventors would be wise to conduct their own preliminary prior art search to find out what examiners might come across during a search so as to assess the chances of success before investing time and money in a patent application. Patent attorneys and specialized patent search companies can assist with doing a search but the cost should be carefully considered given that even the best search will still result in a "maybe." Inventors can also try their hand at doing a search, but finding and analyzing relevant patents and other prior art is often difficult for first-time searchers.

Prior art can be any publicly available product or public technology disclosure. Technology disclosures can be found in formal publications such as scientific or medical journal articles, but can also be from informal sources such as blog posts or YouTube videos. A product itself can also be considered prior art along with catalogs or advertisements for a product. Postings on crowdfunding websites like Kickstarter or Indegogo can also be used as prior art. Although such a wide variety of things can be used as prior art, in the vast majority of examinations patent publications are the only types of prior art that are considered by the examiner.

By default, the contents of a patent application are published while it is pending and then again if it issues as a patent. Both of these patent publications are most often used by examiners because they are easily searched and tend to discuss technology in more depth compared to other types of potential prior art. This is why the term "patent search" is sometimes used interchangeably with "prior art search." A "patent search" is simply a prior art search that is limited to patent publications and does not include a search for the many other types of publications, disclosures, or products that could possibly be used as prior art against a patent application. Although a search for all types of prior art is the most comprehensive, limiting a pre-filing search to only patent publications is often the most cost-effective. In some cases, forgoing a search altogether is the best choice.

Although a patent or prior art search is an integral part of the examination process at the USPTO, applicants are not required to perform a search before filing a patent application. However, if applicants or inventors are aware of any relevant prior art, whether or not it was discovered during a formal search, they must submit this prior art to the USPTO so that it can be considered during examination of the patent application. Failure to disclose known relevant prior art is grounds for invalidation of a patent. Given that performing a prior art search is optional, can have limited value, and even has potential downsides, applicants should carefully weigh the pros and cons before initiating a search.

THE BENEFITS OF DOING A SEARCH

In certain cases, doing a prior art search before filing a patent application can be extremely beneficial. A search can provide a preview of what a patent examiner would find during examination of an application and may even expedite the examination process. Additionally, a search can help prevent invalidation of the patent after it issues and can provide a business with valuable insight into the technology field of its invention before further developing and marketing a product.

BETTER UNDERSTAND THE CHANCES OF PATENTABILITY

The primary benefit of doing a prior art search before filing a patent application is to identify issues that may arise during the examination process. Most importantly, a search may uncover prior art that an examiner would find and use to assert rejections against the application that would ultimately be impossible to overcome. In other words, the time and money spent on a prior art search can be cheap insurance against wasting years of effort and thousands of dollars in fighting for a patent application that will ultimately be unsuccessful.

CAN MAKE EXAMINATION PROCESS EASIER

As discussed in more detail in the coming chapters, the examination process is much like a negotiation. Just as sellers negotiate with buyers to get the highest price for their products, the examination of a patent is effectively a negotiation with the examiner to get the broadest possible protection for a given invention. Seasoned negotiators know that beginning with an aggressive yet justifiable offer allows room for compromises that will allow the opposing side to feel as though it won, without risking an offer that could stifle negotiation by being insulting or signaling lack of sophistication with the item being negotiated for. By knowing the prior art landscape that a patent examiner will encounter during a search, it can be possible to draft a patent application that will be more positively received by the examiner and make a favorable outcome more likely and more likely to occur quickly.

Plus, because the results of a search must be provided to the examiner, it sends a signal that the application was already crafted with the prior art in mind and tends to engender an atmosphere of trust with the examiner. Accordingly, examiners are less likely to overly scrutinize the application and raise needless rejections or unduly limit the protection afforded to the invention where applicants submit prior art references. Moreover, examiners have extremely limited time to examine each application and by effectively doing the examiner's job beforehand or at least making a search easier, the application immediately makes the examiner happy and therefore more likely to treat the application favorably. In fact, in addition to experiencing this first-hand during the examination process, I know several former examiners (now patent attorneys) who admit to favoring patent applicants that would provide even one piece of prior art along with their application. In addition to making the examination process go smoother, performing a prior art search before filing a patent application can result in a stronger patent in some cases.

MAKES A STRONGER ISSUED PATENT

Even after a patent issues, it can later be invalidated by newly discovered prior art in court proceedings or in a reexamination at the USPTO. These proceedings occur most often when a patent is asserted against a competitor. For example, a patent holder files a lawsuit against an infringer and the infringer does a comprehensive search and finds prior art that was missed while the patent was being examined. The infringer can then file a countersuit asserting that they do not infringe because the patent is invalid, or might start a reexamination proceeding at the USPTO so that an examiner can look at the newly discovered prior art and potentially find the patent invalid.

By doing a comprehensive prior art search and submitting the results for consideration during examination, it makes it nearly impossible for this submitted prior art to be used in invalidation proceedings after the patent issues. Conducting a prior art search before filing a patent application can therefore make for a stronger issued patent that is much more difficult to attack and invalidate when it is asserted against infringers.

BETTER UNDERSTAND THE STATE OF THE ART

Aside from the benefits directly related to the patent process and the enforceability of a patent that hopefully issues from the process, conducting a patent search has some benefits for a business in other ways.

For example, searching for old products, existing products, and technology that is under development is a great way to learn about competitors in the field, or potentially identify companies that may want to acquire a license or even buy the company at some point.

Additionally, learning about competitor products and technology can provide insight into mistakes that others have made, deficiencies in existing technology, and unfulfilled consumer needs. Such insights can focus development of a product and even inspire new inventions.

Doing a patent search also reveals how competitors are protecting their inventions and how their patent portfolios are structured. For emerging companies, this can be an important gauge for identifying a patent budget and strategy that will put a company on par with industry leaders.

Although there are numerous benefits to prior art and patent searching as discussed above, there are several important downsides to such searching that makes forgoing a search the default for many companies. At the very least, the practical limitations of doing a search should be well understood before deciding whether to do one.

THE LIMITATIONS OF A SEARCH

One of the most common reasons that inventors want to do a patent search is because they want to know for sure that their patent will be allowed. Unfortunately, the nature of patent and prior art searches makes it impossible to know for sure that an application will ultimately be successful. Because doing a search can never provide the assurances that most applicants are seeking, prior art searching is often not worth the required time and expense.

BEST ANSWER IS ONLY "MAYBE"

By default, pending patent applications are held in secret for 18 months before they are published and become public. Applicants can even elect to keep their applications secret for the entire pendency of the application if they meet certain criteria. Especially in fast-moving fields like computer hardware and software, this means that some of the most relevant prior art will be completely hidden at the time a patent search is performed, but will be available to examiners to reject an application by the time a filed application enters the examination phase. Given that even the best patent search is guaranteed to have huge blind spots, forgoing a patent search is often the best option for certain inventions.

"IS IT PATENTABLE?" IS THE WRONG QUESTION

As discussed in more detail in coming chapters, asking if an invention is patentable is often the wrong question to ask. The reality is that most inventions will ultimately be patentable, but just because a patent issues, does not mean that the patent will have any value. The better question to ask, therefore, is whether patent protection for the invention will have any value. In other words, will the patent provide broad or narrow protection? Unfortunately, the answer to this question is speculative at best, even with the most comprehensive prior art search and intensive analysis by a patent attorney. The value of a patent that may emerge from examination is highly dependent upon which examiner gets the case and how this examiner subjectively determines what prior art is relevant. A prior art search may therefore identify some references that might be considered relevant to an examiner, but cannot anticipate the often unpredictable nature of the patent examination process.

IMPOSSIBLE TO FIND EVERYTHING

Another practical limitation on prior art searching is that it is impossible to find all relevant references, even if they are available at the time the search is performed. For example, recall the story from Chapter 4 of a PhD thesis that was only searchable and discoverable through a card catalog in a special section of a university library. In addition to illustrating the small threshold required for a document to be considered public, this story also illustrates how difficult it can be to find relevant prior art that might ultimately cause a patent application to fail or even be used to invalidate a patent after it issues. This is especially true for publications, advertisements, catalogs, or pamphlets that are not searchable in a database, but still possibly usable as prior art.

The Internet has substantially leveled the playing field when it comes to searching publicly available databases or webpages, but even a good and seemingly exhaustive search can still miss important prior art. This becomes most evident during patent litigation, where hundreds of thousands, if not millions, of dollars are at stake and little expense or time is spared to hunt down even one document or product that might invalidate the patent at issue. In my experience, relevant prior art references can almost always be found, especially when cost is a minor factor.

CERTAIN INVENTIONS CAN BE HARD TO SEARCH

All fields of technology are not equal when it comes to prior art searches, which makes the type of invention a relevant factor when considering whether to perform a search. Physical products tend to be substantially easier to conduct a search for because patent drawings or pictures of a product are much easier to compare to the actual or possible configurations of a tangible device. On the other hand, inventions that are defined by steps of a method, such as software, webpages, smartphone apps, and manufacturing processes, are substantially more difficult to search for because patent drawings or other images rarely provide enough information to make a proper analysis and comparison to the searched invention. For example, a software invention may have certain protectable functionalities, but how these functionalities are achieved through receiving, sending, and processing certain data is what actually defines a patentable invention. When searching prior art references, these specific steps are often difficult to identify unless each potentially relevant document is carefully analyzed. How the steps of a program are described can be extremely varied and inconsistent in the literature, which makes keyword searching ineffective for many inventions. Additionally, the USPTO examination process for physical products is almost always limited to patent publications, whereas the examination of software and other method-based inventions increasingly includes diverse prior art from a wide variety of sources that include webpages, academic publications, and published product specifications. A prior art search for such inventions therefore requires casting a much larger net for relevant material and requires substantially more time to analyze the materials that are ultimately discovered.

Given that complexity and difficulty directly translates into more time and money being required for a search and because the results can never provide the assurance and clarity that many companies often want, choosing to skip a prior art search is often a preferable choice for these types of inventions.

RATIONALE FOR NOT DOING A SEARCH

The main reason that most inventors choose to forgo the optional step of patent searching is because the cost of the search substantially outweighs the relative value gained by doing a search. As discussed in detail below, doing your own patent search is a good way to start, but a patent attorney should always be consulted to evaluate the results and should often be engaged to do some additional searching. For complex and time-consuming prior art searches, the cost of a patent attorney searching or hiring a specialized search firm can rival the cost of just filing a patent application. Since prior art searching is best used as insurance against filing a patent application that is guaranteed to fail, a search is a poor use of resources when just filing a patent application costs less than or even a bit more than a comprehensive prior art search would. This is especially true given how subjective the examination process is and that prior art searches can never guarantee success. More often than not, filing a patent application and letting the examiner do a search and frame the issues is the most cost-effective solution.

In fact, directly filing patent applications instead of first searching for prior art is the default for nearly all companies that have sophisticated patent programs. These companies recognize that having a portfolio of pending patent applications is extremely valuable, even if a few of them end up not making it. Companies that regularly develop technology and file patents want to protect as much of an invention as possible and realize that making even a reasonable estimate of the amount of protection available for each of their applications would be speculative, at best, and not worth the time and cost of prior art researching.

Companies with large research and development programs have patent acquisition models that are much like venture capital (VC) firms that invest in a large number of businesses knowing that most will fail, some will make a bit of profit, and a small minority will be a huge success. These VC firms earn the bulk of their profits from the few investments that end up being massive hits. A stable of companies where the vast majority are guaranteed to be losers or simply average is a central part of their business plan because consistently picking home-run companies is impossible, even for seasoned investors. Similarly, identifying the seeds of a billion-dollar patent at the time of invention is also impossible, even for top patent attorneys with unlimited budgets. Accordingly, by filing a large number of applications, the chances are higher that one or even a few of the patents that emerge will be extremely valuable.

Unfortunately, startups and emerging companies must be substantially more selective about the patents that they file, but should still keep in mind that most established companies do not make prior art searching part of the patent process. That being said, I still highly recommend, and sometimes almost insist on patent searching for certain inventions and with certain types of inventors. For example, with simple products or ideas where the inventors are unfamiliar with the field of technology and where they would likely not move forward with developing a business around a product if it is unpatentable, I will typically recommend that they let me spend one to four hours on patent searching and analysis. This timeframe provides enough time to do an adequate search, but within a reasonable budget that does not rival the cost of drafting and filing a patent application.

Such inventors will frequently do their own patent search (often at my recommendation), but I still strongly recommend allowing me to use their results as a jumping-off point for my own search and analysis, unless their search identifies prior art that would clearly prevent a patent from being allowed.

DOING A SEARCH YOURSELF

Trying your hand at patent or prior art searching can be a great learning experience and is a good way to potentially save some money early on in the patent process. As discussed later in this chapter, there are many excellent free resources that can be used to effectively search through patent literature as well as other types of prior art. By doing a preliminary search before hiring a patent attorney to do one, companies have the opportunity to find knockout pieces of prior art without having to pay a cent.

However, there are a few pitfalls that should be avoided when doing an independent search. For example, for inexperienced searchers, it can be difficult to know which documents are relevant to patentability and how close other inventions need to be for an examiner to reject a patent application for lacking novelty or being obvious. This can cause a couple of problems if the search results are not analyzed by a patent attorney.

First, prior art that comes up in a search might seem to render an invention unpatentable to the untrained eye, but would actually pose no issue when considered by an examiner. Unfortunately, I have informally spoken with many people who have abandoned good inventions or potential businesses based on their own interpretation of prior art search results that a patent examiner would not have looked twice at.

On the other hand, I have also worked with several clients who initially cleared their inventions as being patentable after doing their own prior art search, and wanted to move ahead with filing a patent application without any further analysis or prior art searching. On some occasions, after I insisted on a quick review of the results, it became immediately clear that the invention would almost certainly be unpatentable. In other cases, where a bit of additional attorney patent searching seemed warranted, and after a bit of convincing, less than an hour of patent searching uncovered multiple publications that clearly rendered the invention unpatentable. I hate giving people the bad news that their ideas and inventions will not be patentable but, at the same time, I would feel worse if they had invested thousands of dollars and years of effort in a patent application that never had a chance of success.

Although prior art searching is a great idea for inventors who have the time and inclination, the results should always be reviewed by a patent attorney before making final decisions regarding whether a patent application should or should not be filed. Such a review is typically not time consuming and can often be accomplished during a free initial consultation. Patent attorney review should therefore be the next step after any independent prior art search and additional patent attorney searching should be considered as a follow-up.

PATENT ATTORNEY SEARCHING CONSIDERATIONS

If patent attorney searching is warranted, the scope and budget should be carefully defined before the searching project begins. Despite the benefits of prior art searching, its inherent limitations make it hard to justify devoting a large budget to a search before filing a patent application. In other cases, the relative benefit gained makes it unreasonable to spend any time or money on prior art searching at all.

This is where the experience of a trusted patent attorney should be leveraged. Choosing to do a search depends on the invention and the business plan for exploiting it. With years of experience working with a variety of businesses and evaluating hundreds of inventions in the same technology area, a seasoned patent attorney is in the best position to determine if a patent search would be worthwhile and, if so, identifying a reasonable budget that maximizes the value of the search.

For the cases where prior art searching makes sense, a few hours of attorney searching typically strikes this balance. Any more begins to rival the cost of drafting and filing an application, whereas a smaller search time does not give sufficient time to find relevant results. Additionally, patent searching alone tends to provide the best value for most inventions—especially for physical and mechanical products. For example, a good way to structure a search is to authorize up to three or four hours of patent searching and analysis, with instructions to stop sooner if crippling patent references are found early in the search. A bit of nonpatent prior art searching may be warranted in some cases, but often not more than a half-hour of a three- to four-hour search.

Regardless, the purpose of any prior art searching should be to determine if there are major prior art documents that will clearly make an invention unpatentable, and not to attempt to put a percentage on the likely success of a subsequent patent application. Additionally, the purpose of the search should be to determine patentability and not to determine if a given product or invention might infringe the patents of another.

PATENTABILITY VS. INFRINGEMENT

As discussed above, patentability is a determination of whether a claimed invention is new and nonobvious in view of prior art, which can include patent publications and other technology disclosures or products. The patentability of an invention is decided by a patent examiner at the USPTO during the examination stage of a patent application, and in certain circumstances can be decided in a court of law. For example, granted patents can later be invalidated by a court and rejections of an application by an examiner can also be challenged in court.

Infringement of a patent is completely different and is, instead, a determination of whether a given product or process includes all of the elements defined in the claims of an issued and enforceable patent. A determination of infringement is only decided by courts of law. The USPTO cannot make determinations of infringement. The role of the Patent Office is limited exclusively to assessing the patentability of patent applications and, in some cases, the validity of patents that have already issued. The elements of patent claims and how they relate to infringement are discussed in great detail in the coming chapters but, for now, it is enough to know that determining whether products or processes will infringe a patent is an extremely complex and expensive process.

For many new clients that I work with, in addition to wanting to protect their innovations, they also want some reassurance that their invention or product will not infringe other patents. Such an analysis is called a "freedom-to-operate" opinion and first requires a comprehensive patent search to identify issued and unexpired patents that might have claims that read on the subject technology. For each relevant patent found, a detailed study of the patent must be performed, including a study of the publicly available record of the examination process, including the documents exchanged between the USPTO and the applicant. A chart must then be prepared comparing each claim of the patent to the subject product or invention. In some cases, additional research must be done to determine the validity of these identified patents, which includes a comprehensive patent search on the patent in question. A detailed report is then drafted that formalizes the results of the analysis and advises on whether the company has the ability to practice an invention without infringing any patents.

It should be immediately clear that this freedom-to-operate analysis is substantially more complicated and more expensive than a simple patent search. For example, a simple freedom-to-operate analysis can cost many tens of thousands of dollars. More complicated versions will easily reach into six figures. Determining whether products or inventions might infringe on any patents at such an early stage is therefore extremely rare because the high cost of such an analysis overwhelmingly outweighs the value gained.

In addition to being unjustifiably expensive, the results are also extremely speculative for inventions that are still being developed. Infringement is often highly dependent on specific details of how a product is configured or operates, and slight changes can make the difference between infringement and noninfringement. Therefore, while the exact specifications of an invention are still in flux, it makes even less sense to start considering infringement issues given the high cost and low value of the results. Moreover, even if a product would infringe a given patent, the patent owner would still need to discover the infringing product and invest sizable capital resources in enforcing the patent, which will be done in view of the potential damages that can be obtained and/or the benefit of market exclusivity. Infringing a patent therefore does not guarantee that the infringing party will ever be sued for infringement.

For companies with general concerns about infringement, the best advice is to ignore the specter of patent litigation until a genuine and specific threat is identified. In many ways, being a target of patent litigation is a good thing for an emerging company, because it is an indication that someone considers them to be a serious competitor and one worth spending precious time and money attacking. With many startups struggling to gain the attention of investors and potential customers, it is a sign of success when competitors take note of a company and its products to the degree that they consider patent litigation. Moreover, patent suits are often one step in a negotiation process that can lead to strategic partnerships, cross licensing of IP, and even acquisition of the company being sued.

New companies should therefore focus their precious time and energy on developing the best products possible and bringing them to market, not on the theoretical possibility that they will infringe a patent and be sued. By focusing an IP budget on developing and creating innovative products and investing in a strong patent portfolio, emerging companies can create assets that make them less attractive litigation targets because they have patents of their own that can be asserted.

PATENT AND PRIOR ART SEARCHING RESOURCES

A comprehensive guide to patent and prior art searching is beyond the scope of this book, especially since it varies based on the type of invention being searched. However, the following is an introduction to some of the best free resources available.

USPTO WEBSITE (WWW.USPTO.GOV)

The USPTO keeps a database of all issued patents since 1790, with patents up to 1976 being searchable only by issue date, patent number, and patent classification. Patents issued since 1976 are full-text searchable. All patent publications of pending applications that have occurred since 2001 are text-searchable in a separate database. In addition to searching by patent or publication number, advanced searches based on keyword and classification are also available. Searches can also be limited based on named inventors, patent owners, and even cities where the parties are based. Searches can also be focused on certain parts of an application including the title, abstract, specification, and claims. The USPTO interface is decent, but viewing patent drawings tends to be time consuming, and a separate search must be performed for issued patents and published patent applications, which is quite inefficient. Luckily, there are other free patent searching resources that have substantially better search interfaces with enhanced capabilities.

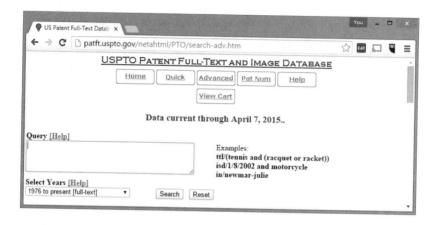

FIGURE 5.1   Screen Shot from Patent Searching Page at USPTO.gov

GOOGLE PATENTS (PATENTS.GOOGLE.COM)

Unlike the USPTO, Google provides for searching both issued patents and patent application publications simultaneously and also provides text-searching of these resources back to 1790. Additionally, the Google patent search includes international publications of issued patents and patent applications from around the world, which are also relevant to determining whether an invention in the United States would be patentable. Searching is lightning fast compared to the USPTO databases, and the interfaces for viewing patent documents is better than at the USPTO. For example, the drawings are easy to flip through, which is essential for certain types of searches. Plus, for issued patents, Google lists all patent documents that were considered relevant by the examiner during examination of the application and even lists later patents where the examiner found that patent relevant to a determination of patentability. This feature effectively allows a searcher to leverage the searches of USPTO examiners, which makes it possible to find leads on patent documents that might otherwise be difficult to search.

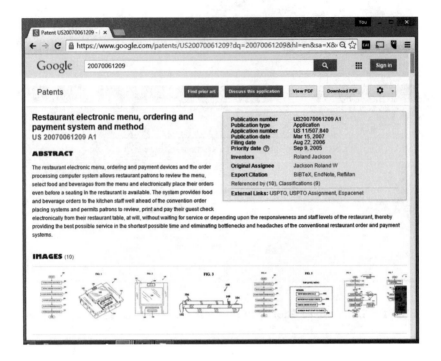

FIGURE 5.2   Screen Shot of a Patent Displayed by Google Patents

However, Google Patents does not currently include the robust searching by invention classification that is available at the USPTO. Although classification searching turns up a lot of irrelevant leads for most inventions, it can be essential in some searches and should not be completely overlooked. For searching patent application publications and publications of issued patents, Google Patents is an amazing primary search resource, with the USPTO searches being a good way of finding material that a Google search might have overlooked.

GENERAL WEB SEARCH

In addition to searching for patent publications, searching for other types of prior art can be important for a well-rounded search and is a good way to fill in the gaps left by the unavailability of patent applications that are held in secret for 18 months or more from filing. By doing keyword searches for webpages, publications, images, videos, and existing products, it is possible to discover critical prior art. A Google web search can be a good way to search for such prior art, but Google has some specific search tools that are also quite useful. For example, Google Scholar (scholar.google.com) and Google Books (books.google.com) allow for text searching within books, academic publications, and other formal technology disclosures. Google also has specific search engines for shopping, images, and videos, which can be a direct source or lead on relevant prior art.

FINAL THOUGHTS ON PRIOR ART SEARCHING

Because it is an optional step in the patent process, applicants should assume that no patent or prior art search is being conducted by their patent attorney unless it is explicitly discussed before drafting a patent application begins. When clients do not broach the subject, most patent attorneys will not actively recommend prior art searching unless they believe the facts appear to weigh in favor of it. Accordingly, where there is concern about patentability or a question about whether a prior art search is warranted, inventors should speak up and discuss the issue.

Although searching is often not a cost-effective option for many inventions or companies, having a patent attorney walk through the rationale for performing or not performing a search in a given case is still a useful exercise. Since the decision to conduct a prior art search is highly subjective and based on the type of invention, budget, and business goals of the applicants, taking the time to discuss the case-specific merits of patent searching is necessary to make a well informed assessment.

Regardless, if applicants have the time and inclination, doing a preliminary search is often helpful. In addition to potentially finding prior art that might be relevant to patentability, a prior art search is a great way to better understand the landscape of competing products and businesses, which is an important insight for any growing business. With the numerous free resources available for patent and prior art searching as discussed in this chapter, there is relatively little down-side to conducting an initial search before engaging a patent attorney to review the results. For more information on patent and prior art searching, including up-to-date insider techniques for efficiently find-ing relevant prior art with the searching resources discussed above, visit www.PatentsDemysitified.com.

CHAPTER 5 SUMMARY

- Prior art can be any publicly available product or technology disclosure (e.g., live presentations, patent publications, journal articles, webpages, or videos).
- Patent or prior art searching is *not* a required or default step before filing a patent application.
- Searching can help determine if an invention is unpatentable or can make the examination process at the USPTO easier.
- On the other hand, searching often provides limited value for the cost, especially given that prior art patent applications are held secret by the USPTO for at least 18 months from their filing and may therefore be unsearchable.
- Applicants can try searching on their own, but should have the results analyzed by a patent attorney.
- Relevant prior art that was already known or discovered in a search must be submitted to the USPTO when filing a nonprovisional patent application.

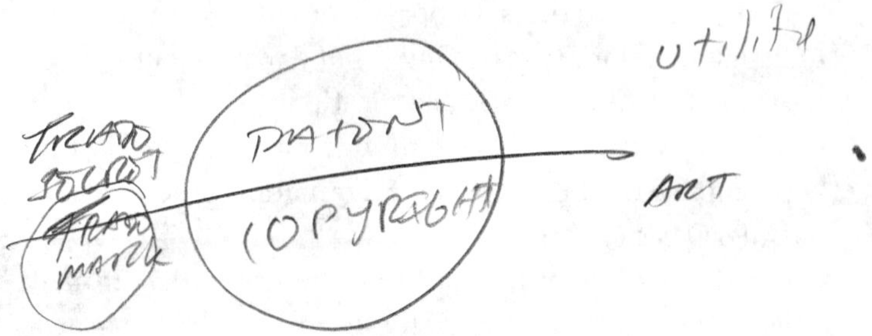

TRIAD
SECRET

TRADE
MARK

DIA TONY

COPYRIGHT

utility

ART

6

Choosing What Type of Application to File

"Whoever invents or discovers any new and useful process, machine, manufacture, or composition of matter, or any new and useful improvement thereof, may obtain a patent therefor, subject to the conditions and requirements of this title."

—Title 35 of the United States Code, Section 101

As discussed in previous chapters, the most common type of patent application is a utility application. However, applicants can choose to file a provisional or nonprovisional utility application as a start to the utility patent process. Choosing which one is a matter of cost, how developed an invention is, the type of invention, and the importance of having enforceable rights quickly. U.S. Patents only provide protection within the bounds of the United States. Foreign protection is obtainable, but must be secured separately in each country where a patent is desired. This chapter explores the options of starting the patent process with a provisional or nonprovisional utility application and then discusses foreign patents along with the other types of U.S. patent applications.

PROVISIONAL VS. NON-PROVISIONAL UTILITY APPLICATIONS

COST

Provisional applications incur less cost upfront, but end up increasing the total cost of the patent process. As discussed in previous chapters, a nonprovisional must be filed within a year of filing a first provisional, and while material from the initial provisional application can often be reused in the nonprovisional application, there is additional cost compared to only drafting and filing a nonprovisional as a start to the process. For example, at the very least, there is additional time in filing an extra application and additional USPTO filing fees. However, any downside of total overall cost by using a provisional application strategy is often outweighed by numerous potential benefits that provisional patent applications provide.

DESIRE TO DEFER COSTS

Many start-ups and individual inventors choose to file provisional applications, despite the increased overall cost, because of a desire to defer the cost of nonprovisional applications and the examination phase. One common reason for deferring costs is to test the market before investing capital in the drafting and examination of a nonprovisional application. Additionally, many companies want to find investors or other sources of capital to pay for building a patent portfolio at the nonprovisional stage and want to invest the least amount possible upfront before such financing is secured.

Having a year to test the market can be important to determine whether the cost of obtaining a patent is a wise business decision. For example, if a product fails to sell well within its first year after filing a provisional application, and the company decides to discontinue it, having filed the provisional application instead of a nonprovisional saves the applicant several thousand dollars.

Similarly, a year after filing a provisional application, product price and projected sales figures may indicate that the time to recoup the costs of filing and examination of a patent application does not justify the benefit of pursuing a patent. Again, thousands of dollars can be saved if a provisional application is abandoned instead of a nonprovisional application.

Also, applicants seeking capital investors or other financing are able to delay the costs of drafting and examination of a nonprovisional, with the intention of having another party eventually pay for these expenses. Moreover, already having a provisional application filed makes investing in a project substantially more attractive.

For these reasons, a provisional patent application is used as a cost-effective way to get a foot in the door of the patent process without incurring substantial upfront costs. This flexibility can make the overall increased cost worthwhile.

PROGRESS OF INVENTION DEVELOPMENT

Another reason for filing a provisional application as a beginning to the patent process is to allow for changes and advances in technology during product development. For example, inventors are wise to file a patent application shortly after conception of an invention, but inventions are far from static, especially in these initial stages. In some cases, an invention will evolve through dozens of iterations as it develops into a marketable product. A final commercial embodiment may end up being completely different from an initial prototype or concept.

In such situations, provisional patent applications offer a cost-effective way of protecting various stages of product development. In fact, a common strategy is to file a first provisional application and, within the year that it is pending, file one or more additional provisional applications to capture filing dates at important development stages. At the year anniversary of the first provisional application (when it is set to expire) the progression of provisional applications are combined, along with any further developments, into a nonprovisional application. An example of this strategy is shown below in Figure 6.1.

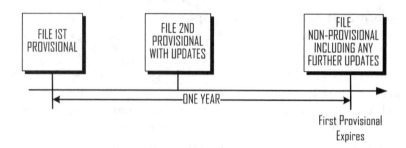

FIGURE 6.1

While it is possible and often necessary to capture steps in a development process once a nonprovisional application has been filed, doing so is substantially less economical and creates a series of active patent applications that each require considerable attention and extra incurred cost. Accordingly, when a product development process is anticipated, a provisional patent application can be an ideal first step.

APPLICATION PREPARATION TIME

It takes a substantial amount of time to properly draft a nonprovisional patent application. Unfortunately, inventors and companies do not always allow sufficient time for drafting a nonprovisional application, which should ideally be allocated several weeks. Luckily, the loose requirements of provisional applications provide an acceptable alternative when last-minute or emergency applications are needed. As mentioned in previous chapters, foreign patent rights are often lost upon a first public disclosure, public use, or offer-for-sale, with U.S. patent rights being lost a year after such occurrences. Many times, an important disclosure or sale opportunity will arise before applicants expect them, and passing it up would not be a wise business decision. At the same time, it would also be unwise to make such a sale or disclosure without first filing a patent application.

When situations like this arise and there is not sufficient time to draft a top-quality nonprovisional application, a provisional application can provide much needed protection at the last minute if absolutely necessary. After the disclosure or sale opportunity, a nonprovisional may be filed soon afterward or can be filed on the year anniversary as normal. Also, if quality of the provisional was compromised due to severe time constraints, a second comprehensive provisional could be drafted and filed later after adequate preparation time. However, this multiprovisional strategy does incur more cost compared to filing a single provisional application.

NEED FOR ENFORCEABLE RIGHTS

Having an issued patent is required to have enforceable patent rights and filing a provisional application effectively adds a year to the time it takes to get an issued patent, assuming the nonprovisional is filed exactly at the one-year expiration date. For this reason, applicants might prefer nonprovisional applications when faced with current or imminent infringement, or when licensing or sale agreements provide an incentive for having an issued patent sooner.

TYPE OF TECHNOLOGY—VALUE OF A PATENT OVER ITS TERM

In addition to delaying issuance of a patent by a year, filing a provisional application to begin the patent process also effectively shifts the term of a patent forward a year because patent term is calculated as 20 years from the filing date of a nonprovisional application.

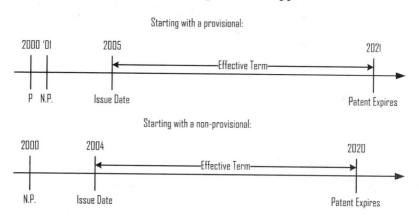

FIGURE 6.2

For example, Figure 6.2 depicts how the term of a patent filed in the year 2000 would be shifted by a year depending on whether a provisional or nonprovisional patent application was filed first. In both scenarios, the first application is filed in the year 2000; pendency of both applications is assumed to last three years; and the term of both patents is 20 years from the filing date of the nonprovisional application. The active term of both patents is also the same, namely 15 years.

The important difference between these two hypothetical patents is the choice of having an enforceable patent during the period of 2004 to 2005 or during the period of 2020 to 2021. While having an active patent shifted a year earlier or later may not seem like a big difference, it can become extremely important when considering the projected value of the patent during these two periods. For example, consider Figure 6.3.

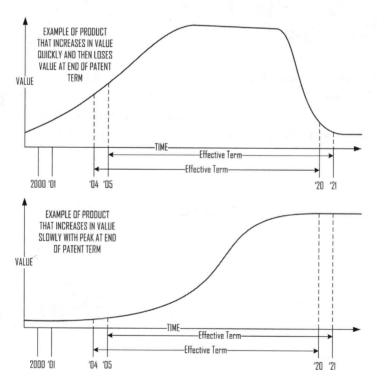

FIGURE 6.3

In both figures, the curve represents value of the patent or technology over time. The first figure illustrates technology that increases in value rapidly, peaks, and then quickly loses value. For example, computer hardware, software, and Internet-based technology can have a value curve like this first graph because development of such products can occur relatively quickly and will also be adopted by the market relatively quickly. However, within a few years, because of how quickly computer and Internet technology advances, this same technology also becomes rapidly outdated.

Such a value profile makes the earliest years of a patent term more valuable than the last years. The dashed vertical lines depict the shift in patent term based on whether a provisional or nonprovisional application is used to begin the patent process. Here, the area under the curve between the dashed lines is equal to total value for that year period. Because the value between 2004 and 2005 is at least three times greater than the value between 2020 and 2021, filing a nonprovisional and obtaining an issued patent faster can increase profit and value in this first example.

On the other hand, the second figure shows a value profile that makes the later years of a patent term substantially more valuable than the beginning years. In this case, filing a provisional application to delay the issuance of the patent by a year will greatly increase the total value of the patent over time. In fact, choosing a provisional application adds a year of patent term that is approximately ten times as valuable.

This second value profile can be typical of pharmaceuticals and biotechnology. Unlike computer-related technology, developing medicines and therapeutics typically takes a long time and is extremely expensive, which cuts into the value of such technology in the initial term as this cost is recouped. Extended research and development time may be due to product trials or the lengthy FDA approval process. However, once the product is ready for market, and development costs have been recouped, such products reach and hold their peak value with only a few years of patent term remaining.

While it is impossible to accurately predict how a given invention's value will change over time, understanding how products typically perform over a couple of decades can help you decide whether to file a provisional or nonprovisional patent application. Accordingly, predicting a value profile for an invention can be an important step when deciding how to begin the patent process.

REAL-WORLD EXAMPLES

With various factors introduced for choosing either a provisional or nonprovisional application as a start to the patent process, some real-world examples are useful in illustrating how these factors are weighed based on the inventors, business plan, and technology at hand. The following examples provide an analysis of theoretical, yet common, fact patterns that are typical of potential clients that I meet with on a regular basis.

EXAMPLE 1

Ryan is a recently retired auto worker who invented a bottle-opener attachment for a television remote control. He took an old opener he had laying around the house and attached it to a wooden mock-up of a device that would snap onto an average TV controller. As a next step, he wants to hire a product designer to help him refine a version that could be mass produced, but wants to protect his invention before talking with the designer.

Here, Ryan should probably file a provisional application to begin the patent process. Given that he is an individual inventor, he might have a limited budget that should be conserved for initial expenses such as product development and production along with business expenses such as marketing. Also, because the product is in initial stages of development, a provisional application would give Ryan a year to refine his product and possibly make changes to the design before filing a nonprovisional. Finally, the provisional application gives him the opportunity to test the market for a year without incurring the full costs of a nonprovisional application.

EXAMPLE 2

Jen and Kevin work as project managers at Macrofirm Software but are starting their own side business by pooling the $20,000 bonus that *they each received this year. They are developing LicensePlateDate.com, which is a dating website where drivers register their license plate as a username, and can communicate with other users by text, phone, and email by using the license plate number of drivers they see on the road. A user can receive alerts when potential date matches are driving near them and locate them via GPS. Jen and Kevin plan on starting a test market in San Francisco and then expanding nationwide with the help of venture capital investors.*

Although Jen and Kevin could afford to file a nonprovisional patent application, they are still likely better off filing a provisional application first because of where they are in the development process. As new functionalities are invented in the first year, updated provisional applications can be filed with relatively little additional cost to protect these new inventions. Even with a decent start-up budget, they would be wise to conserve their capital early on as they work to build the website and grow a critical mass of users. Also, in case their business ends up not working out, they will have invested less in patent costs up front. Moreover, if they are able to find investors within the first year, their nonprovisional patent costs can be covered by someone else's money.

EXAMPLE 3

Davin is the CEO of Fun-Lectric Inc., a small personal electronics company that has been in business for nearly five years. Until now, the company has sold products made by other companies or reproductions of common home electronics. However, their engineering team recently came up with the "Phone Radar," which is a novel handheld device that locates active cell phones in a room. Ten-thousand units have already been ordered and the product will be marketed to schools and airlines where cell phone use is prohibited. Davin wants to file a patent application before Fun-Lectric begins selling and promoting this product.

In this case, Davin and Fun-Lectric should file a nonprovisional to begin the patent process. The product has already been developed and produced and is ready to be sold, so there is little reason to conserve money for purposes of marketing or further development. In fact, starting with a nonprovisional application will save on patent costs over time. Also, because the product is already in commercial form, there is less reason to file a provisional application in anticipation of changes in the product.

EXAMPLE 4

Dawn and Doug were on a road trip in Oregon and almost got into an accident when a car in front of them slammed on its brakes without warning. On the way home, they thought up a way of preventing car crashes by creating an LED brake light matrix that would illuminate more rows of lights based on how much brake pressure was being applied by the driver, and blink when maximum pressure was being applied. In the following weeks, they sketched out the details of how their brake light invention would work and even managed to line up a meeting with the director of a traffic safety advocacy organization who they hoped would make their brake light a federal standard. With this meeting only a week away, they realized that they should probably file a patent application first to protect their invention.

Fortunately, despite having less than a week to get a patent application filed, Dawn and Doug will still be able to protect their brake light idea before this important meeting. Inventors should try to budget several weeks to find a patent attorney and get a good provisional application filed, so a week is going to be a rush job. A nonprovisional can take a month or more to draft and it is a bad idea to rush a nonprovisional. For this reason, filing a provisional is the best option because of the more lenient requirements of provisional applications.

Also, once they have established an initial filing date with a first provisional application, and had their meeting, Dawn and Doug may be wise to file another provisional application if the quality of the first application was compromised due to the rush. Alternatively, they may decide to file a nonprovisional soon after if their situation changes.

EXAMPLE 5

Brian invented a new engine supercharger, which he disclosed to several attendees at a meeting of a local entrepreneurial networking group. Unfortunately, a few weeks later, when doing some market research on his invention, he discovered the blog of a company that was announcing the release of a product that was nearly identical to his invention.

Upon further investigation, he recognized one of the company directors as being present at the meeting where Brian presented his idea. Brian confronted the president of this company about unfairly using his ideas, but the president informed Brian that talking about the idea publicly was essentially the same as giving others permission to use the idea.

The president is absolutely wrong. Although public disclosures can harm patent rights, and are strongly recommended against before filing a first patent application, the one-year grace period in United States likely gives inventors a year to file a patent application after making a public disclosure before patent rights are completely lost. Brian might consider filing a nonprovisional patent application because infringement of his idea appears to be imminent. Because Brian will not actually have enforceable rights until his patent issues, he needs do everything possible to get it issued as fast as possible. A provisional application would effectively delay issuance by a year, so starting the patent process with a nonprovisional might be preferable in this situation. However, because priority is given to the first inventor to file, Brian should make sure to file as soon as possible in hopes that he files before the competitor does. Accordingly, filing an emergency provisional followed quickly by a nonprovisional application may also be a good choice.

As a side note, be wary of being intimidated by people who appear to be important, successful, or experienced, especially when their interests are opposing yours. In the example above, the company president tries to convince Brian that there is no issue because Brian does not have any enforceable rights. Maybe the president is intentionally misleading Brian, or perhaps the president does not know the law. Regardless, even if the president seems to know what he is talking about, Brian needs to do his own due diligence and determine what the law is.

In general, inventors should be wary of patent advice given by anyone who is not a patent attorney. Such free advice might be the most costly advice that one gets because it may result in forfeiting patent rights or inadequate protection at the very least. There are numerous urban myths about how to protect ideas or about how patents work, and I continuously hear them parroted by experienced business people and even people who have a bit of experience with the patent process.

For this reason, always consult a patent attorney before making important decisions about intellectual property.

EXAMPLE 6
Britany's husband Brandon gets hiccups all the time. One day she started experimenting with a mixture of herbs that could be sprayed into his mouth to stop a case of the hiccups. After nearly a year of testing and refining, she completed a final formula for her anti-hiccup spray, and started to work on making it into a product that could be sold in stores. Unfortunately, she discovered that hiccup treatments were governed by the FDA, and that she would have to conduct several years of clinical trials to prove that it was safe for the public.

Even if Britany had the budget to file a nonprovisional application, she should still file a provisional application in this case. Having a fully developed product and wanting to save money in the long run would normally indicate that a nonprovisional would be appropriate, but the ability to shift the effective term of a patent would make filing a provisional application desirable here.

Because it may be several years before Britany can even put her product on the market, her patent would likely issue before her FDA testing was complete, and having a patent during this time would have minimal value. Britany's hiccup spray would likely have a patent value profile much like the one shown earlier in this chapter that peaks late in the term, so she would be capturing more value over time by delaying issuance of her patent by filing a provisional to begin the patent process.

FOREIGN PATENT APPLICATIONS AND THE PCT

As discussed in Chapter 3, U.S. patents are only enforceable within the bounds of the United States. Accordingly, if protection is desired elsewhere, it must be obtained on a country-by-country basis. Foreign protection must be secured within one year of a first U.S. application, regardless of whether this first application is a provisional or nonprovisional application. One option for foreign protection is to file directly in desired countries at the one-year mark. However, another option is to file a PCT application,[1] which allows applicants to efficiently file subsequent foreign patent applications at least 18 months later in most foreign countries. The PCT application does not mature into a patent, and merely acts as a placeholder for later foreign filings in specific jurisdictions. The benefits of the PCT application are to delay the choice of filing in specific countries by at least 18 months and possibly to reduce the overall cost if several foreign applications are ultimately filed. There are a few regional patent offices, such as Europe and Eurasia, but many important markets require filing in an individual country including Japan, China, Canada, Australia, and India.

Although having patents in countries aside from the United States can be important in certain situations, the vast majority of applicants should be wary of filing foreign patent applications. Foreign patents are much more expensive than U.S. patents, and often require yearly maintenance fees while the application is pending and each year the issued patent is in force. In contrast, U.S. patents only have three maintenance fees due after the patent issues. Moreover, foreign markets tend to be much smaller than the United States, and the cost of enforcement in foreign jurisdictions also tends to be more unpredictable and costly than in the United States.

Unless a company already has a large established market and a hefty patent budget, the relative value compared to the cost does not justify pursuing foreign patents. Without overriding justifications, extra patent budget is better spent reinforcing and building a U.S. patent portfolio instead of getting bogged down with foreign protection.

Commonly, companies substantially overestimate their future market penetration in foreign markets and file a slew of expensive foreign patent applications expecting rosy profits in these foreign jurisdictions. Unfortunately, without the proper connections or experience, developing foreign markets can be difficult and ends up sucking energy away from developing primary markets such as the United States. In fact, foreign patent assets are more likely to become a liability to a company than a valuable asset.

---

1. Recall that PCT stands for Patent Cooperation Treaty, which is an international treaty that provides a unified procedure for filing patent applications in each PCT member country.

For example, even as the reality of actual profit potential from foreign countries becomes clear, few companies are willing to cut their losses by dropping these applications and, instead, continue to sink money into foreign applications that have little value compared to other patent assets. Over time, the money spent is unlikely to be recouped in these foreign markets.

However, the ability to obtain foreign patents can be important to certain investors or business partners, so leaving open the potential for foreign protection might still be important. For this reason, companies sometimes choose to file a PCT application so that international "patent pending" status can be claimed, even if they do not intend to pursue foreign patent applications 18 months down the road. Making a company more attractive for acquisition, an IPO, or a round of funding might therefore tilt the scales in favor of filing a PCT application.

OTHER TYPES OF U.S. PATENT APPLICATIONS

Provisional and nonprovisional patent applications are two versions of a utility patent application, which is by far the most common patent application that is filed and issued. However, there are a couple of additional patent species that may come into play in certain limited situations—design patents and plant patents.

Plant patents specifically protect asexually reproduced plant varieties, including cultivated sports, mutants, hybrids, and newly found seedlings, other than a tuber-propagated plant or a plant found in an uncultivated state. Of the nearly half a million patent applications filed yearly, only about 1,000 are plant patents. If you believe that you have an asexually reproduced plant that should be patented, find and consult a patent attorney who has experience with plant patents. However, it is substantially more likely that a plant-related invention will be protectable with a utility patent.

Design patents are substantially more common, with more than 25,000 filed yearly. Recent changes in the law have substantially expanded the protection that design patents afford, which has made design applications increasingly more popular as a way of augmenting a patent portfolio. Design patents cover the ornamental design of an article of manufacture, compared to functional products and methods that are the subject of a utility patent.

Unlike utility patents, design patents provide protection that is limited to a single specific design of a product and do not cover different variations or looks. Moreover, design patents only cover the look of a product and not the underlying functionality of an invention. Nonetheless, a design patent can be a great option when utility patent protection is not applicable or if getting a utility patent would be difficult or unduly expensive. Also, because design applications are relatively inexpensive and tend to issue substantially faster than their utility counterparts, filing a design application in addition to a utility application can be a fast way to get initial patent protection before a utility application issues and often even before the utility application is examined. Filing design and utility applications together can also be highly economical because, if planned correctly, many of the same drawings can be used in both applications.

Design patent protection is most applicable when there is one design, or specific features of a design that makes a product distinctive. The protected designs and features cannot be dictated purely by function however. Also, design patents are only applicable to three-dimensional features of a product, and not to paintings or drawings that may be present on a product. Copyright or trademark protection may be a viable way to protect such two-dimensional artistic features as discussed in Chapter 1.

Design patents are different from utility patents in that there is no such thing as a provisional design application—a nonprovisional design application is the only option. Also, while utility patents have a term that runs 20 years from the nonprovisional filing date, design patents have a term that lasts 14 years from the date that the design application issues as a design patent.

Companies that are considering design patent protection should consult a patent attorney who has experience with drafting and examination of these applications. While design applications are relatively simple in form, maximizing scope of protection can be tricky. Inexperienced patent attorneys tend to overlook simple ways to expand design protection, so be sure to work with an attorney who knows about design patents.

Regardless of whether a given product is best protected by utility or design patents, it is best to engage an attorney for both planning and drafting of any applications. Additionally, although it is possible for inventors to draft their own applications and brave the examination process at the USPTO alone, doing so it not recommended. The following chapter makes the case for why working with a patent attorney is the only viable option for companies that intend to enforce, license, or otherwise exploit their patent assets.

CHAPTER 6 SUMMARY

- Applicants can begin the utility patent process by filing a provisional or nonprovisional application.
- A provisional application can provide a lower-cost option for starting the patent process and also allows for cost-effective updates to the application as the invention changes during the one-year term of the provisional application.
- When having enforceable rights faster is beneficial, and the invention is already solidified, a nonprovisional application to start the patent process can be a good choice.
- Separate foreign applications are required for patent protection outside the United States, but are often undesirable because of the high cost for relatively low value.
- A design patent application can be helpful where protection is needed for a specific ornamental design of a utilitarian product.

Why You Should Work
with a Patent Attorney

*"An examination of this application reveals that applicant
is unfamiliar with patent prosecution procedure. While an
applicant may prosecute the application . . . lack of skill in
this field usually acts as a liability in affording the maximum
protection for the invention disclosed. Applicant is advised to
secure the services of a registered patent attorney or agent to
prosecute the application, since the value of a patent is largely
dependent upon skilled preparation and prosecution."*

—USPTO Manual of Patent Examining Procedure § 401—
Standard USPTO warning to inventors representing themselves.

Most inventors are capable of drafting a patent application that meets
the basic filing requirements set forth by the USPTO. Applicants that
represent themselves, known as pro se applicants, are often able to successfully get issued patents. In fact, many books[1] are available to assist

---

1. For example, *Patent It Yourself: Your Step-by-Step Guide to Filing at the U.S. Patent Office*,
by David Pressman, Nolo Press.

applicants with the drafting and examination phase of their patent applications. Unfortunately, what these guides fail to tell eager inventors is that only a minority of pro se applications actually mature into issued patents, and of the ones that do issue, the vast majority are effectively unenforceable or invalid for various reasons.

What few inventors realize is that all patents are not created equal, and simply getting a patent issued is not a guarantee that it will have any value. Getting a patent issued is relatively easy, but it is extremely difficult to get one issued that maximizes protection of the invention and that will also survive the intense scrutiny that comes during enforcement, licensing, or sale of the patent. This chapter makes the case that serious companies and inventors should always work with a patent attorney, unless they have no intention of actively using their patent assets. Working with a patent attorney should be affordable for most businesses and inventors, and the coming chapters provide insider tips and tricks on how to save money and work cost effectively during the whole patent process without having to sacrifice overall patent quality.

DOWNSIDES OF DOING IT YOURSELF

A recent study[1] found that only 23.6% of pro se applicants were able to get their nonprovisional patent applications granted, in contrast to a 65.2% success rate for represented applications. Pro se applicants were found to have more trouble during examination and received substantially more rejections than attorney-represented applicants. By not realizing rejections are a standard part of the process, not knowing how to respond, or simply becoming frustrated and giving up, only 41.2% of pro se applicants even responded to a first Office Action from the examiner, and only 32.2% responded to a second one. Represented applicants responded 85.6% of the time to a first Office Action and 78.7% to a second. Further data indicated that pro se applicants that

---

1. The Lone Inventor: Low Success Rates and Common Errors Associated with Pro-Se Patent Application, Kate S. Gaudry, PLoS One, Volume 7, Issue 3, e33141, published March 21, 2012, *available at* http://www.plosone.org/.

were able to get an issued patent had patent claims that provided less invention protection compared to represented applicants.

These statistics are indicative of a system that does not favor pro se applicants and, at the same time, only barely scratches the surface of the potential consequences that pro se applicants usually face.

GRANT OF A PATENT THAT CANNOT BE INFRINGED

As indicated in the study above, the minority of pro se applicants that successfully obtain a patent are more likely to receive a patent that provides less protection than one obtained with the assistance of an attorney. More specifically, the patent usually has less value because the wording makes it less likely that competitors will infringe. Understanding how patents can broadly or narrowly cover an invention is one of the most important concepts explored in this book, and is discussed in detail in the coming chapters. For now, however, it is enough to know that patent value is directly tied to how broadly the patent protects an invention, and that pro se applicants tend to unintentionally word their patents such that protection is narrower than it could be.

Wording mistakes not only sometimes reduce value, they can make a patent completely worthless because it is effectively impossible to infringe. For example, in a 2004 case, Chef America, Inc., was suing a competitor for infringing its patent on an innovative method of making delicious pastries from dough. All elements of the infringement case appeared to be met but the patent claims required a step of "heating the resulting batter-coated dough to a temperature in the range of about 400° F to 850° F." At a glance, this seems to be a standard cooking instruction, but notice that the step actually says *heating* the batter-coated dough at 400° F to 850° F, not *baking* the dough in an oven that was heated to 400° F to 850° F. In other words, this element literally requires the dough itself to be heated to 400° F to 850° F, which would result in a burnt, black, and crispy mess that would be completely inedible.

The applicant clearly meant to say "baking" instead of "heating," and it might seem reasonable to infer that correct meaning in a case like this. Unfortunately for Chef America, the court found that, even if "construing the patent to require the dough be heated to 400 degrees to 850 degrees Farenheit [sic] produces a nonsensical result, the court cannot rewrite the claims" of the patent. Therefore, by inadvertently using the word "heating," the patent was effectively worthless because no baker in his right mind would ever use a method that essentially created a burnt piece of coal. Seasoned patent attorneys know how to watch out for traps where a seemingly harmless error can make a patent invalid and, still, mistakes can happen. Pro se inventors are substantially more likely to make fatal mistakes like this that cannot be undone.

SLOWER AND MORE EXPENSIVE EXAMINATION PROCESS

As indicated by the study discussed above, pro se applicants are more likely to have issues complying with USPTO filing formalities, which directly translates to a slower and more expensive examination process. For example, applicants can theoretically go back and forth with an examiner indefinitely, but after every two or three rounds, the applicant must pay additional fees to keep the process going. Examiners have no official time limit for responding to correspondence from applicants, but applicants must respond to Office Actions within two to three months before having to pay extension fees that grow larger every month. However, after six months, and sometimes sooner, failure to respond results in automatic abandonment of the application. Luckily, the USPTO allows applicants to revive applications that accidentally go abandoned but require a hefty fee to do so. Similarly, when other deadlines or formalities are not observed, the USPTO often provides for a fix, but again, a sizable fee is typically required.

I see these problems often when new clients ask me to take over for them in the middle of the examination process. The time, effort, and fees required to fix the application to give it a fighting chance can rival the cost of having a patent attorney work on the application from the beginning and, still, a resulting patent will typically be less valuable. Saving money by drafting and filing a patent application without a patent attorney can therefore cost more money in the long run while also resulting in an inferior issued patent.

GRANT OF AN INVALID PATENT

Unfortunately, patents can issue with hidden defects that ultimately render them invalid. These problems lie dormant until opposing parties have an opportunity to scrutinize the patent and pick it apart, such as when the patent is enforced, licensed, or sold. Patent attorneys, and especially patent litigators, are keenly aware of how to find and exploit these weaknesses in the patents of others because it can mean that their client will be free of liability from infringement or can justify paying substantially less or nothing at all when buying or licensing a patent. Patent attorneys that know how to destroy patents are also great at constructing patents because they can anticipate and defend against attacks that might not occur until years in the future.

Pro se applicants are unable to anticipate these weaknesses while drafting and during examination of a patent application because the issues arise from areas that are never checked or questioned by the USPTO. For example, as discussed in Chapter 5 on prior art searching, the inventors and applicants have an affirmative duty to disclose any relevant prior art that they are aware of; failure to do so is considered fraud against the patent office and can result in invalidation of the patent. Companies and inventors are often completely unaware of this requirement and do not learn about it until it is too late.

Patents can also be invalidated after they are issued if opposing parties discover public uses, public disclosures, or offers-for-sale that occurred more than one year before the priority date of the patent. These types of activities are not investigated by or even discoverable by examiners aside from all but the rarest cases. However, these harmful activities almost always come to light in the spotlight of litigation or due diligence. Again, pro se applicants are often unaware of the rules and consequences of these activities until it is too late.

These are only some examples of the hidden pitfalls that lurk in the shadows of the patent process and are all but obscured to unsuspecting applicants that file their own applications without the assistance of a patent attorney. Unfortunately, parties that are experienced with patents appreciate the potential liabilities that come with self-drafted patent assets and use this to their advantage.

PRO SE PATENTS ARE LESS ATTRACTIVE TO INVESTORS, PARTNERS, LICENSORS, AND BUYERS

Self-drafted and examined patents are discriminated against by patent attorneys and others that are sophisticated in working with patent assets. These parties know that such patents are more likely to have less intrinsic value or be completely worthless because of hidden invalidity issues. Moreover, when money or potential infringement is on the line, these parties will take the time to study the complete public filing history, including all correspondence with the USPTO, to specifically identify the flaws lurking in the patent's past.

Having a self-drafted or examined patent can therefore be a hindrance when trying to attract investors or business partners because it signals that the company is under-capitalized or does not take intellectual property protection seriously. It also indicates potential weakness in the patent assets and therefore justifies lower valuation of a company. With investors and collaborators being extremely selective about with whom they work, it is not uncommon for deals to fall through because of deficiencies in patent assets.

Similarly, during negotiations for the sale of a company or its patents, or the licensing of technology, the fact that patent assets were self-drafted or examined is frequently used to negotiate a lower price for the buyer or licensor. In other cases, buyers will decide to back out completely or potential licensors will decide that a license is unnecessary because the patent does not sufficiently cover desired technology or is otherwise unenforceable because of defects.

For example, when I represent clients looking to buy, license, or invest time or money in patent assets of another company, one of the first things I look at is whether any part of the patent history shows signs of a self-drafted or examined application. Where this is the case, and because it is a huge red flag, I recommend that the company not consider working with the patent asset until a detailed evaluation is completed. For some clients, this immediately makes them move on while others agree to the cost of a detailed review and analysis, but only if the patent owner accepts a lower price to offset this cost and potential risk.

Fortunately, working with a patent attorney should not be prohibitively expensive for most businesses and the coming chapters provide the tools needed to cost effectively grow a patent portfolio without wasting time and money. With the negative consequences of forging through the patent process unrepresented in mind, the remainder of the chapter explores the benefits that patent attorneys provide above and beyond their basic duty of drafting, filing, and prosecuting patent applications.

BENEFITS OF HAVING AN ATTORNEY DO IT

As discussed above, working with a patent attorney makes it more likely for patent applications to grant as issued patents and results in assets that are perceived as, and actually are, more valuable and enforceable. However, working with a trusted patent attorney has numerous additional benefits.

A good patent attorney can be an important advisor and ally for businesses, especially for new and emerging companies. Because many patent attorneys work with technology companies of all sizes on a regular basis, they are familiar with what makes technology companies more likely to succeed or fail. The positive and negative experiences of their clients provide important insight into growing and maintaining technology companies in addition to just patent strategy.

For example, by working with cutting-edge technology on a daily basis, patent attorneys are often aware of trends in various technology sectors. Although they may not have experience with your specific invention, they may have insight into how other technologies will interface with, augment, or improve your invention. In fact, when drafting patent applications, it is not uncommon for the process to be a catalyst for ideas that completely change the direction of product development or inspire new embodiments of an invention that the inventors had not originally contemplated. All patent attorneys are required to have a technical background, and many also have independent business experience or have worked in an industry where they at one time were inventors and product developers themselves. As discussed in more detail in Chapter 8, you should investigate and ask about the unique skills that your patent attorney may have, and leverage these extra talents when possible.

This unique experience and familiarity with technology companies of various sizes directly informs patent work as well, and allows a patent attorney to formulate a custom patent strategy that fits the individual technology, business plan, and budget for each company. Having a tailored strategy is important because it ensures that capital is not wasted on unnecessary patent protection or that important inventions do not go unknowingly unprotected. More often than not, working with a patent attorney generates and saves substantially more value than would have been saved if patents had been self-drafted and examined.

Moreover, working with a patent attorney provides assurance that all important deadlines are being carefully monitored, that unknown pitfalls are being avoided, and that the filing and examination of an application is being handled in the most efficient way possible. Inventors that draft and examine their own applications often spend large amounts of time doing so and spend even more time worrying about whether they did everything correctly and did not overlook anything. They tend to spend needless time and money battling with the USPTO simply because they do not know the rules and culture of patent examination. By offloading this time and effort, these inventors are able to devote their valuable time and energy in developing their invention and growing a profitable business. Such undivided attention by founders and inventors is typically essential to the success of a company, and lack of focus is a common reason many businesses fail.

As discussed above, self-drafted and examined patents are often negatively viewed, which can result in lost profits and missed opportunities. On the other hand, working with the right patent attorney or firm can serve to open important doors and attract important business partners and investors alike, while at the same time intimidating potential infringers.

For example, companies that are seeking investors and business partners use a business plan as a marketing tool when approaching these parties. One of the most important portions is the biographies of the founders, key employees, and any advisors. People who have experience working with start-ups know that even the best business idea is worthless unless there is a team that can successfully implement it. Therefore, by being able to state that a company has engaged a top firm or well-respected patent attorney serves to show that the business team has experience and guidance necessary to succeed, which is especially important when the founders are not seasoned entrepreneurs.

Selecting the right patent attorney is therefore critical not only to the success of a patent portfolio, but also to the health and future of a growing company. Qualities such as experience, technical background, and firm type are just some of the many factors to be considered. The following chapter teaches not only how to find a great patent attorney, but also how to find a patent attorney that best fits your unique business goals, budget, and technology.

CHAPTER 7 SUMMARY

- All patents are not created equal—they can be broad or narrow and can also have hidden defects that make them effectively worthless.
- Accordingly, the purpose of the patent process is not simply to obtain an issued patent, but to obtain one that is as broad as possible and free of defects that would make it unenforceable.
- Applicants that draft and file their own applications or self-navigate the examination process are substantially more likely to never obtain an issued patent, and even if they do, these patents will likely have less value and be difficult to enforce, license, or sell.
- Given that working with a patent attorney is possible on nearly any reasonable budget, serious applicants should not attempt to obtain patents on their own.

8

Picking a Patent Attorney

"No individual will be registered to practice before the Office unless he or she . . . (i) Possesses good moral character and reputation; (ii) Possesses the legal, scientific, and technical qualifications necessary for him or her to render applicants valuable service; and (iii) Is competent to advise and assist patent applicants in the presentation and prosecution of their applications before the Office."

—Title 37 of the U.S. Code of Federal Regulations, Section 11.7

A patent attorney should be carefully selected, just as any business partner or advisor would be selected. Patent attorneys do more than just draft and file patent applications for you—they should be integral and active advisors in planning and executing a personalized intellectual property strategy for your business. The right patent attorney can add great value to a company, whereas the wrong one will waste valuable capital resources that are best used elsewhere. Finding a trusted patent attorney who understands your business and invention and has experience crafting and executing successful patent portfolios is therefore essential.

Patent attorneys must have special qualifications to properly guide clients in the complex landscape of patent law. Obtaining a patent requires submitting an application with the USPTO, and anyone who represents clients before the USPTO must pass the "patent bar," which is a qualifying exam (much like a state bar exam for attorneys) that establishes competency in patent law and in the mechanics of drafting, filing, and examination of patent applications at the USPTO.

Additionally, given that patents relate to technology, the USPTO also requires that any person sitting for the patent bar exam must have at least an undergraduate degree in a technical or scientific field or sufficient equivalent experience. For example, a degree in electrical engineering, computer science, or biochemistry qualifies an individual to sit for the patent bar exam. While this is the minimum required educational background, it is not uncommon for patent attorneys to have advanced degrees and even experience in industry or academia.

Although the vast majority of patent professionals are attorneys, there is not a requirement to be an attorney to represent clients at the USPTO. Patent bar members who are not attorneys are known as "patent agents."

In the eyes of the USPTO, patent attorneys and agents enjoy equivalent rights in terms of representing clients during the filing and examination process. However, the main distinction between patent agents and patent attorneys is that agents are not licensed to practice law, and are therefore unable to perform even the most basic legal tasks, some of which can be important to the patent process. For example, a patent agent cannot draft legal documents, such as a license, a nondisclosure agreement, or an assignment of patent rights, and may be limited to the use of attorney-approved forms. Additionally, patent agents cannot make legal determinations related to patent infringement because such an analysis requires extensive knowledge of federal patent law.

Despite an inability to perform legal services, patent agents should not necessarily be overlooked when choosing a professional to work with—their inability to practice law is one of many considerations to weigh. However, given that the majority of patent bar members are attorneys, the following discussion relates primarily to patent attorneys.

CHECKING CREDENTIALS OF PATENT ATTORNEYS

After passing the patent bar, each new attorney or agent is issued a "registration number" or "reg. number." Registration numbers are issued sequentially regardless of agent or attorney status, so a registration number can give you an idea of a given professional's seniority. Before engaging anyone to help you with a patent, make sure that this person has a registration number with the USPTO. Lacking a registration number makes someone ineligible and unqualified to represent clients before the USPTO and should automatically disqualify this person as a potential representative.

Unfortunately, I have seen many people harmed by attorneys and organizations that claim, sometimes with the best of intentions, that they are qualified to draft and file patent applications. For example, one of my favorite clients lost the ability to protect an important invention after receiving help from a well-intentioned business attorney friend. Before being referred to me, this client had been using this attorney for years to help him with business affairs. Most likely, this business attorney wanted to keep his long-time client happy, but unfortunately miscalculated the complexity of patent law and was unaware of the requirement to be patent-bar registered. While attorneys are licensed to practice nearly any sort of law that they want within certain jurisdictions, patent law is one of the few exceptions. Although the business attorney discussed above managed to get a provisional application filed with the USPTO, the application provided essentially no protection, and any ability to patent the invention was subsequently lost because the business attorney did not realize that the provisional application would automatically expire at the end of its one-year term.

The easiest way to confirm that an attorney is a patent attorney is to look online at www.uspto.gov. As an exercise, try finding my profile within the USPTO database by searching my full name, or even simply searching for "Dylan," which returns fewer than ten entries. From there, it is easy to find my registration number, which is 56,289.

FIGURE 8.1   Screen Shot of an Avvo.com Attorney Profile Page

In addition to checking your attorney's status at the USPTO, you may also want to make sure that he has a clean record with the local bar association. Avvo.com is a great resource for getting background information about attorneys practicing any type of law, including patent law. Avvo attorney profiles, at the very least, include information about bar admission status and whether an attorney has ever been disciplined by a bar association. Additionally, the bar association website from the state in which the attorney is licensed in can be a useful tool as well. For updated links to patent attorney searching resources, along with a free downloadable form for evaluating and comparing different patent attorneys, visit www.PatentsDemystified.com.

WORK DONE BY PATENT ATTORNEYS

The primary job of a patent attorney is to represent clients in the patent process, which includes planning a patent strategy, drafting and filing patent applications, and working patent applications through the examination process. When the option of prior art searching is chosen before filing a patent application, patent attorneys can perform the search and analyze the results.

Patent attorneys are often able to perform other patent-related services, such as analyzing products to determine if infringement of an issued patent is likely present, analyzing patents to determine if they might be invalid, and litigation in court to assert or defend against claims of patent infringement. However, patent litigators are not always patent attorneys and there is no special technical background requirement like the patent bar for attorneys to represent clients in federal court for patent matters. Patent attorneys typically do make excellent patent litigators because of their experience with patent drafting and the examination process, but nonpatent attorneys should not be ruled out when it comes to asserting patents in court or making infringement determinations. Assisting with reviewing contracts for patent licensing or sale of patents can also be done by a patent attorney, but again, this is also a service that general attorneys can handle.

Because patents are in the family of "intellectual property" along with trademarks, copyrights, and trade secrets, patent attorneys will also typically have some general knowledge about these methods of protection. However, such so-called "soft IP" is rarely a major part of a patent attorney's practice, and so it is often better to find a specialist in these areas when necessary. Additionally, patent attorney rates are typically higher than nonpatent attorneys, so finding a specialist in trademarks, copyrights, or trade secrets is also typically a more economical option.

This rationale also applies to other legal and business services. For example, some patent attorneys might have experience incorporating a business, developing inventions, planning business strategy, or obtaining business financing, but companies should carefully consider whether the relatively high billing rate of a patent attorney is worth it for services that might be performed less expensively by nonpatent attorneys or even nonattorneys.

On the other hand, experienced patent attorneys can have invaluable personal or professional experience in such business or legal areas, which can be worthwhile to tap into on a limited basis. For example, patent attorneys who have worked closely with a variety of clients over time, or who have personal business experience, can be excellent sources of business strategies that separate successful companies from those that fail. Especially for new and growing companies, patent attorneys can provide valuable insight into the steps that can take a company to the next level. Similarly, because they work with cutting-edge technology on a daily basis, have a technical background, and often have hands-on experience as engineers, scientists, or other technicians, patent attorneys can also be a great resource for product development ideas and understanding landscape of a given technology market. Even with their relatively high billable rate, patent attorneys can be excellent resources on a limited consulting basis.

However, for time-intensive services that are outside their typical job description, the high rates of patent attorneys are likely not justifiable for services such as searching for a buyer or licensee for a patent and securing financing or investors for a company. Finding investors or buyers is a time-consuming job and can be very specialized depending on the industry or type of product that is involved. For example, brokering a sale for the rights to kitchen products is entirely different from brokering a sale of the rights to a website, and the buyers or licensors are going to be entirely different types of companies. Brokers in these fields are highly specialized because they know the right people and also know how to approach them. The practice of patent law rarely overlaps with such a specialized skill set, and paying a patent attorney by the hour to acquire these skills is not an efficient use of a budget.

That being said, patent attorneys in specialized niche technology may have some helpful insight. For example, one of my colleagues has extensive experience with high-grade imaging crystals that are produced by only a few companies worldwide. He is familiar with each of these companies and knows all the experts in the field who study these crystals, along with the few companies that would buy such technology.

However, even if you happen to find patent attorneys with such specialized knowledge of your product or invention, their insight may actually prevent them from being able to help you because doing so would create a conflict of interest with their existing clients and may pose a risk of exposing attorney-client secrets.

Regardless, do not expect your patent attorney to be a primary source of finding parties to buy, license, or invest in your technology. Aside from being a potential source for advice in this arena and helping review contracts in such a deal, a patent attorney should not be paid to act as a broker. A trusted patent attorney may be able to provide a referral in certain cases, but would not likely be the best person to actively perform the work.

Additionally, patent attorneys rarely invest in a company by performing services in exchange for stock or otherwise deferring costs until a company becomes profitable. Like most businesses, the practice of law has continuing overhead costs that include office space, staff, and the like. Unfortunately, this means that long-term deferred payment plans are not viable options for the vast majority of patent attorneys who need a steady stream of income to support their business.

One exception is when a law firm is representing a company in a financing transaction. Corporate attorneys doing such a deal might defer some of the fees for such a client and a deferred-fee structure might also be extended to other legal services, including drafting and filing patent applications. However, in such cases, the law firm and company being represented are both fairly well established, and the primary services are financing or corporate legal services. This, among many other factors, might impact a decision to choose one attorney over another.

PROFESSIONAL SETTINGS OF PATENT ATTORNEYS

Patent attorneys who can represent new clients at the USPTO are only found in a couple of professional settings—working alone or within a firm. Although patent attorneys work in a multitude of other settings, inventors should limit their search to those in a firm or solo practice.

Patent attorneys who work alone are known as solo practitioners. These attorneys might work from home, have their own offices, or may work in a shared office space that includes other attorneys, professionals, or businesses, but the solo attorney will be a separate business entity. Solo patent attorneys may or may not have personal staff, such as a secretary or paralegal. On average, solo patent attorneys will have a lower billing rate than patent attorneys working in a firm. Additionally, solo patent attorneys typically work with individual inventors and small- to medium-sized businesses, but in some cases may also represent larger clients.

A firm of less than 100 attorneys or that specializes in a specific area of law is typically known as a "boutique firm." Some boutique firms may only have a single patent attorney among a group of attorneys, while other boutique firms consist primarily of patent attorneys.

Patent attorneys typically find the most synergy with nonpatent attorneys who have a business, litigation, or intellectual property focus, but there is no limit to the variety of law practices that may reside within a firm where patent attorneys might practice. For boutique firms of only patent attorneys, patent-related work typically dominates the focus of the firm, which may or may not include patent litigation. However, even if a boutique firm includes only patent attorneys, it will still usually advertise itself as an "intellectual property firm" instead of as a "patent firm." Boutique firms will almost always have staff to support the attorneys and may even have offices in multiple locations. Clients of boutique firms are typically of all sizes, ranging from individual inventors to large multi-national companies.

Firms that employ more than 100 attorneys are generally considered mid-sized to large firms. These firms typically offer a wide variety of legal services with patent attorneys working in a patent or intellectual property group within the firm. Large firms typically have an extensive support staff that works alongside the attorneys. Although large firms typically cater to large companies, they are capable of representing clients of any size.

Patent attorneys can also be employees at a technology company or university, where they represent the company in a variety of patent and business matters. Also known as working "in-house," these patent attorneys typically collaborate with external solo attorneys or firm attorneys that represent the company in legal matters. Such patent attorneys are not able to represent clients outside the company in which they work and should be ignored when searching for patent counsel. Accordingly, solo attorneys and firms are the only choices for finding an eligible patent attorney with which to work.

HOW TO CHOOSE A PATENT ATTORNEY

Choosing the right patent attorney is an important decision when trying to get high-quality patents while optimizing a limited patent budget. With billing rates that vary from $100 per hour to $1,000 per hour, it can be tricky to pick an attorney who provides the most value without being cost-prohibitive. Good patents do not come cheap, but more expensive is not always better, especially when considering the type of technology, available capital, and purpose for obtaining a patent. The sections below discuss the most important factors to consider when searching for a patent attorney.

TECHNOLOGY EXPERIENCE

During the process of drafting patent applications, a patent attorney must quickly become an expert on the invention that is being protected, which requires that he or she understands the technology as well, if not better, than the inventors. Finding an attorney with the right technical background is therefore essential. Patent attorneys are generally specialized in two areas—computer technology and biological technology. This differentiation is due to the technical language of the biological sciences being drastically different from the language of computer software and hardware. Moreover, the style of drafting patent applications is very different for these two families of technology.

Accordingly, inventors of computer-related inventions should ideally seek out patent attorneys who have at least an undergraduate degree in electrical engineering or computer science, and experience with the drafting and examination of computer-related patent applications. Specific experience with hardware or software may also be beneficial, depending on the type of invention. On the other hand, patent attorneys who work in the biological sciences tend to be PhDs and will work with a wide variety of technology, with inventions that relate to humans, animals, plants, and microbes.

Few patent attorneys focus exclusively on mechanical devices even though mechanical engineers are eligible to become patent attorneys. Inventors with these types of inventions should be open to patent attorneys with nearly any technical background, but with sufficient experience in drafting and examination of mechanical devices.

Finding a patent attorney with an educational or professional background in a specific technological sub-discipline can be beneficial, but is typically not necessary or possible. Seasoned patent attorneys are experts at quickly mastering inventions in a general technology area and crafting them into patent applications that maximize protection while conforming to patent laws. On balance, finding a patent attorney that has extensive experience with a specific type of technology is much less important than many of the other factors discussed below.

Patent attorneys' educational and industry experience is relatively easy to surmise from their firm or personal website, but it is not always easy to tell how much experience the attorney has with the drafting and examination of certain types of technology, and often requires asking the patent attorney directly.

PERSONAL ATTENTION TO YOUR CASE

One way to improve the chances of getting a well-drafted patent application at the lowest possible cost is to find a patent attorney who feels personally invested in your company. In other words, it is best to work with an attorney who is enthusiastic and passionate about the company's success and the technology being developed. Although good attorneys always strive to do the best for each of their clients, an attorney who is emotionally invested in a client tends to give extra attention and will potentially go easy on billable hours where possible. As discussed above, it is rare to find a patent attorney who will invest in a company by taking stock or other profit-sharing stake, so any personal investment tends to be of the emotional variety.

Having a relationship with a patent attorney is one way to get a personal investment, but a direct relationship is not always necessary.

For example, a referral from an attorney's existing client or friend can also be a good motivator. In these cases, he or she will want to provide quality service so as not to disappoint a friend or potentially harm the relationship with a good client.

However, the best way to guarantee extra attention as a client is to find a patent attorney who is genuinely excited about your technology and company. Patent attorneys tend to be technophiles and naturally have favorite types of technology or inventions. For example, they may have studied a given technology in school or in industry, or may simply have a personal interest in certain types of products or technology. Alternatively, attorneys may have an affinity for certain types of companies or business plans. For example, an attorney may specifically enjoy working with graduate students who are spinning off companies from a university or may like the energy of tech startups that are pushing for an IPO.

This type of personal connection assures that extra care and attention will be given to your case, which typically includes what I call "shower time." In other words, your patent attorney is so engrossed in your work that he thinks about it in the shower, on the way to work, at lunch, and at other personal times. In addition to not typically being billable time, this extra thought translates into a better planned patent portfolio, higher quality patent applications, and more comprehensive analysis during the examination process. Accordingly, finding a patent attorney who has a personal connection with you, your company, and your technology can be extremely valuable.

Unfortunately, some patent attorneys or firms do discriminate against smaller clients. The business rationale is that it is less risky and takes less time on average to maintain relationships with fewer larger clients that are better capitalized and have a larger volume of work than it is to maintain relationships with many smaller clients that will file fewer patent applications. T However, these patent attorneys and firms will not necessarily decline representation of smaller clients, which is can be extremely undesirable for these unknowing companies who often receive less personal attention and care than they deserve. To avoid falling into this trap, always ask patent attorneys about the types of clients that they typically represent and directly ask them about their preferred size of clients.

Another pitfall to avoid is having your work passed on to an uninterested patent attorney. In firms with multiple attorneys, it is not uncommon for a senior attorney to enlist the help of a junior attorney in drafting patent applications or drafting responses to Office Actions from the USPTO. This junior patent attorney will do the majority of the work, often under supervision of the senior attorney. This can be beneficial in keeping costs down because the junior attorney typically has a lower billing rate, but this can be problematic if the junior attorney is inexperienced or is not excited about working on your patent application. Accordingly, when choosing a patent attorney, a primary consideration is who will actually be doing the work on your patent application. For example, it is not uncommon for a senior attorney to give lower-priority patent work to law clerks (i.e., law school interns) or new associate attorneys for training purposes. Even with good supervision, there is a greater possibility of a lower-quality work product overall.

Additionally, where work is handed off between patent attorneys, there is no guarantee that the same attorney who drafted a patent application will be the attorney who handles the examination process. In fact, it is not uncommon for different attorneys to handle different stages of the examination process. This is a distinct disadvantage for companies that want more than a cookie-cutter patent portfolio, because, having one or more new attorneys pick up a patent application cold, without background knowledge of the invention and company, makes it more likely that the final work product will implement a patent strategy that does not fit the unique needs of the applicant.

On the other hand, working with junior patent attorneys is not always a bad thing. In addition to typically having a lower billing rate, a junior attorney may have even more relevant technology background and be more enthusiastic to do top-notch patent work than the senior attorney who delegated the application. Therefore, when choosing a patent attorney, it is important to ask who will be doing the bulk of the work, and make a determination of whether this person is sufficiently qualified and motivated to work on your case. The time invested in finding a patent attorney with the perfect background and skill set is worthless if this ideal attorney hands your case off to someone who is a poor fit.

EXPERIENCE

When hiring a patent attorney, determining relative experience is much more complex than the number of years a patent attorney has been practicing or the date that the attorney passed the patent bar. Instead, the number of years of *relevant* patent and technology experience is the best gauge of how qualified a given patent attorney is to handle the unique patent needs of your company.

In terms of legal experience, an ideal candidate is a patent attorney who has many years of experience with drafting patent applications and working these applications through the examination process. Moreover, this should be the core of the patent attorney's practice—at least 80 percent or more. Mastering the skills of planning patent portfolios, drafting patent applications, and negotiating the examination process requires consistent focus over time, and specialists are far superior to attorneys who dabble in the patent application practice. For example, some attorneys will take the time to pass the patent bar because they are eligible, but rarely have any meaningful direct experience with the patent process. These attorneys may instead focus on business formation, trademarks, copyrights, trade secrets, licensing, litigation, or any number other legal specialties. Accordingly, the number of years that attorneys have been registered patent attorneys is meaningless unless their practice has been almost exclusively focused on patent applications.

Knowing how to enforce patents and defend against patent invalidation challenges makes patent attorneys substantially better at handling patent applications because this experience directly translates into knowing how to build patent assets that are easier to enforce and less likely to be invalidated in court. Therefore, a top candidate is a patent attorney who also has patent litigation experience but that still has a practice that is devoted to patent application work at least 80 percent of the time.

For technology experience, it is important to not only consider a patent attorney's educational background and hands-on technical experience, but also the attorney's experience with patent applications of a given technology type. For example, if a patent attorney has a degree in computer science, but only has experience with patent applications related to mechanical devices, this attorney would likely be a less-desirable choice for working on a patent related to software or computer hardware. Accordingly, in addition to considering the technical background, you should also determine how much of a patent attorney's practice includes patent applications that are in the same field as your invention.

A third consideration is a patent attorney's experience working with clients that are similar to you and your company. For example, a patent attorney who only has experience with large clients that have hundreds of patent assets would likely be a poor choice for small or even mid-sized companies that are often more cost-sensitive and in need of a more focused patent strategy. On the other hand, patent attorneys who have experience only with individual inventors or small businesses would not likely be a good choice for mid-sized or rapidly growing companies due to limited experience with building larger patent portfolios or crafting strategies for emerging companies. For small and mid-sized growing businesses, an ideal candidate will have worked with a wide variety of clients, which would provide insight into building patent assets that fit the current needs and budget of the company, while also crafting a strategy that will grow and adapt as the business expands.

HOURLY BILLING RATE/PROJECT RATE

A patent attorney's hourly billing rate should be considered in view of relevant experience as discussed above. More relevant experience should justify a higher billing rate. Hiring top-notch patent counsel does tend to be more expensive, but seasoned patent attorneys are typically able to be more efficient and are therefore capable of doing more work in less time. For example, the cost for a patent attorney with a billing rate that is double that of another patent attorney will likely be more, but probably substantially less than double. Therefore, the best way to compare the relative cost between attorneys is to compare a projected total project cost instead of simply hourly billing rates.

LOCATION

Unlike the practice of some other types of law, assisting clients in obtaining a U.S. patent is not limited to a geographic location. U.S. patent attorneys can be found in most countries around the world, and these attorneys can represent clients before the USPTO regardless of where they are located. For example, as a licensed Washington state attorney and a registered patent attorney, I have represented patent clients based in most states and many foreign countries, including Japan, China, Italy, Australia, England, and even Cuba.

Despite being able to engage a patent attorney located in most countries around the world, there should be an overriding reason for choosing one who is not local. Any part of the patent process can be accomplished without face-to-face contact between attorney and client, but having the ability to collaborate in-person can certainly improve communication and improve the quality of the attorney's work. There are, however, a few good reasons to consider a remote patent attorney.

One reason to seek out a remote patent attorney is because the local selection is limited or unsuitable. Patent attorneys tend to be more concentrated in major metropolitan areas—especially cities with technology-based economies. Accordingly, applicants that are far from major markets might want to expand their search to encompass a wider selection of patent professionals. Local patent attorneys may simply not have suitable experience or seasoning. On the other hand, for companies based in expensive markets like New York or San Francisco, it may be desirable to seek out less expensive, yet equally experienced patent counsel in another area.

FINDING AND RESEARCHING PATENT ATTORNEYS AND FIRMS

Finding a comprehensive and relevant list of patent attorneys or law firms is surprisingly difficult. The USPTO offers a searchable online database of all registered patent agents and attorneys by geographical area, but the information about each registered individual is limited to only their name, location, and company affiliation, if any. Inactive attorneys, in-house attorneys, and patent attorneys practicing other types of law are all listed together, so many listed are patent attorneys who cannot help you or are undesirable candidates. Additionally, there is no way to limit a search by technology practice area.

Using conventional methods of Internet searching is also unwieldy and often provides a limited selection. Listings such as online directories tend to be unreliable because firms must typically pay to be listed, and the majority of patent or intellectual property firms or attorneys do not pay for any formal marketing at all. If you do find online listings of patent attorneys or firms, keep in mind that such a search method is terribly limited.

One of the better online attorney search portals is Avvo.com. Although this website includes a listing of attorneys with all practices, it does have good search functions that can narrow a search to find relevant patent attorney results. Most attorney profiles include a ranking and details about a given attorney's background and practice along with notification of any ethics complaints or issues.

However, top patent attorneys typically only get new business from word-of-mouth referrals from existing clients or other attorneys. In fact, if you ask patent attorneys where they get their best new clients from, they will emphatically tell you that referrals are their preferred method of meeting new clients. For this reason, the most reliable way of finding good patent attorneys or firms is to tap your network and ask around. This can be especially important for finding a patent attorney who has some expertise in a given technology. If you do not have anyone in your network who has experience with patents, members of local entrepreneurial or technology groups can be a valuable resource.

After identifying firms or patent attorneys through networking or online resources, always check patent attorney profiles on their personal or firm web page. This profile will normally highlight an attorney's educational and professional background, including relevant work experience, along with representative technology with which the attorney commonly works. Browsing a firm's website will also give you an impression of what a firm's focus is, whether it is a specific type of legal practice, a certain type of client, or the firm's core values.

With one or more patent attorney selected, contact them directly to see if they seem like good matches. When making first contact via phone or preferably e-mail, give a little background about yourself, how you heard of the attorney, and some basic information about the invention you want to protect. Unless asked, avoid getting into excessive detail during this first contact. Focus on introducing yourself and picking a mutually convenient time to meet if you feel the potential for a good fit. Always schedule an initial consultation with an attorney in advance. If you are interested in a particular firm, you should still identify an individual attorney at the firm with whom you want to meet. Otherwise, you will simply be paired with the first patent attorney that happens to be available.

If you already have information prepared about your invention, such as descriptions, drawings, or pictures, offer to forward these materials to the attorney before you meet. Although you should not always expect attorneys to thoroughly study and analyze such materials, giving them time to review and understand an invention beforehand will make an initial meeting substantially more productive.

However, there is no need to prepare special documents, disclosures, or prototypes before meeting with a patent attorney. Now that the United States gives priority to the first inventor to file a patent application, unnecessarily delaying the patent process can mean the difference between a patent application being allowed and being rejected. While drawings, descriptions, or prototypes can be helpful, they are not necessary for patent attorneys to understand an invention and draft a patent application, and applicants that agonize over perfecting these items and wait to seek patent help are doing themselves more harm than good.

THE INITIAL CONSULTATION

After making a timely arrival at an initial meeting, it is best to start with the basics. Begin with a brief personal background and discuss personal experience with the technology field of the invention. Introduce the history of your existing company or the vision of a company that will soon be created.

Most importantly, discuss how patent and other intellectual property protection will fit into a business plan. These details provide important context that will help you and your patent attorney craft a patent strategy that serves the unique needs and goals of the growing company.

When discussing your invention, you may consider using visual aids or even bringing in a prototype if possible. Most attorneys will have a computer and a whiteboard in their meeting space, so consider using these tools as well. In addition to disclosing your preferred or ideal version of the invention, talk about alternative ways to achieve the same desired effect, even if they are less desirable. Similarly, it is helpful to also disclose what future versions of the invention are expected to be after additional research and development.

Also discuss whether any public disclosures, uses, or offers-for-sale of the invention have been made. A "public" disclosure in the legal sense can be to a single person, so a patent attorney should be informed of anyone outside of the company who knows about the invention. If any such disclosures, uses or offers-for-sale were made awhile ago, it may be important to determine the exact date that they occurred or at least a specific date range.

After a general disclosure of the invention, the next step is to begin sketching out a patent plan. Should you file a provisional or nonprovisional application? Is a patent search desirable and, if so, how detailed should it be? What patent application strategies should be employed? With a course of action plotted, the cost of the project should be determined. For example, a projected cost range or maximum cost should be agreed upon for each stage of the patent process. The cost of drafting and filing the application can often be estimated and quoted, but a cost for the examination process cannot because this stage is so unpredictable. However, be sure to completely understand the costs associated with drafting and filing, and then to get projected costs for best-case and worst-case of the examination stage. Along with getting cost estimates, an agreement should be made about any advance fee deposit or retainer that needs to be paid before work begins.

For firms that include more than one attorney, a conflict check must be run before any party is officially accepted as a client of the firm. Conflict checking refers to the process of cross-checking a potential new client against a database of existing firm clients to determine if this new client would create a conflict of interest with existing firm clients. For example, a firm should not represent two clients that are adverse or potentially adverse to each other in litigation and should not represent two clients that have extremely similar inventions. For larger firms, completing a conflict check can take a couple of days, whereas a solo attorney will immediately be able to assess any conflicts at an initial meeting.

Once a new client passes the conflict check, the attorney-client relationship is formalized with an engagement letter that spells out the terms and scope of the patent attorney's representation. It should address issues such a billing, scope of representation, and the like. Along with signing an engagement letter, the new client will also likely be asked to provide a retainer or advance fee-deposit before the patent attorney begins work on the project. Typical amounts of such an advance are the full estimated cost of the project, or a large portion of the estimated project cost. At this point, the patent attorney has officially been hired and drafting of a patent application or a preliminary prior art search can begin.

CHAPTER 8 SUMMARY

- Only registered patent attorneys and agents are allowed to represent clients before the USPTO.
- The primary job of a patent attorney is handling the patent application process, which includes planning a patent strategy, drafting and filing patent applications, and working patent applications through examination at the USPTO.
- Applicants should seek out a patent attorney with a technical background that matches their invention and who also has experience working with similar companies.
- Years of focused patent application and relevant technology experience is a good gauge of how qualified patent attorneys are.
- Applicants should hire a patent attorney who is not only skilled in the mechanics of the patent application process, but one that will also be a trusted advisor in crafting and implementing a customized patent strategy throughout the life of the business.

Drafting an Invention Disclosure

"Invention is not enough. Tesla invented the electric power we use, but he struggled to get it out to people. You have to combine both things: invention and innovation focus, plus the company that can commercialize things and get them to people."

—Larry Page, co-founder and CEO of Google

After choosing a patent attorney, many inventors are unsure of how to properly and efficiently disclose their invention so that a patent application can be drafted. My new clients are sometimes dissatisfied when I ask them to simply give me any documents, descriptions, diagrams, or pictures that they already have so that I can sort through them to determine what is relevant and what additional information might be needed. They can still be uncomfortable even when we discuss a follow-up inventor interview to fill in any gaps. Instead, many inventors would prefer coaching about what documents are relevant and how to draft additional documents that describe their invention in the "correct" way.

For many reasons, working with existing documents and conducting an inventor interview is the most efficient way to obtain the relevant information that will form a foundation for an attorney-drafted patent application. First and foremost, teaching someone how to draft a perfect invention disclosure that is relevant to a specific invention often takes substantially more attorney time and incurs more cost than necessary. Additionally, drafting formal invention disclosure documents is often a daunting and intimidating task that can take weeks or even months to complete. Along with being a poor investment of time and energy, delaying the filing of a patent application and not establishing the earliest filing date possible is often an unacceptable risk.

On the other hand, taking the time to write an invention disclosure can be an extremely valuable exercise if done during the development of an invention and at the early stages of a business. By having to articulate an invention in writing, inventors can gain new insights into their inventions, which aids in the research and development process. Additionally, invention disclosures are often a starting point for a business plan and other promotional materials that are essential to any growing business.

Invention disclosures can also form the basis for provisional patent applications, and inventors, along with some patent attorneys, will sometimes file them directly as provisional patent applications. Using an invention disclosure as a provisional application can be a great strategy in certain emergency situations, but harbors risks that inventors and even some patent attorneys fail to fully recognize. Accordingly, before going into detail on drafting a comprehensive invention disclosure, it is important to understand why having a quality provisional patent application is so important.

WHY PATENT ATTORNEYS SHOULD IDEALLY DRAFT PROVISIONAL PATENT APPLICATIONS

The USPTO will accept essentially any document submitted as a provisional patent application. This is a double-edged sword because, while filing a provisional application provides the opportunity to submit informal documents, it also creates the opportunity to submit documents that provide no protection and actually result in the loss of patent rights.

YOU ONLY GET PRIORITY FOR WHAT IS SUBMITTED

As discussed in previous chapters, the purpose of a provisional patent application is to establish an early priority date for a subsequent non-provisional patent application that must be filed within one year. This early priority date should prevent loss of patent rights due to public disclosures, uses, and offers-for-sale, while also gaining priority over later patent applications filed by others and various types of later prior art.

However, provisional patent applications are not substantively examined by the USPTO, which effectively means that any document can be filed and granted a filing number and priority date. This ability to file any document and have it accepted as a provisional application has created a common misconception that any document filed as a provisional will give protection to an invention, regardless of the content and quality of the disclosure. To the contrary, a provisional application is still held to the standard of allowing one of ordinary skill in the art to make and use the invention that is later claimed in the nonprovisional patent application. A provisional application that fails to meet this standard is effectively worthless because it does not provide the benefit of an early priority date. Moreover, the illusory protection provided by such a provisional patent application often results in unintentional forfeiture of patent rights.

For example, in 1996 New Railhead Manufacturing LLC began selling a new drill that provided for controlled horizontal and angled drilling into hard rock formations, which created increased efficiency and accuracy when boring railroad tunnels and the like. The success of its invention prompted New Railroad to file a provisional patent application on the invention in 1997 and subsequently file a pair of nonprovisional patent applications later that year, which claimed the benefit of this provisional patent application. Both applications issued in 1999 and claimed a drill head that was angled in relation to the drill body. However, when these patents were later asserted against competitors, the defendants searched through the publicly available patent file history and discovered that this special angled relation of the drill head and body was not adequately described in the provisional patent application. The court agreed and held that the issued patents could therefore not receive the benefit of the early priority date of the provisional application.[1]

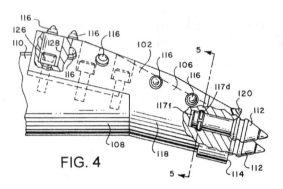

FIG. 4

FIGURE 9.1   **Drawing of a Drill Bit for Horizontal Directional Drilling from U.S Patent 5,899,283**

Unfortunately, without the priority of the provisional application to save them, sales of the drill made in 1996 invalidated the issued patents because these sales occurred over one year before the nonprovisional applications were filed. It was undisputed that New Railroad was well aware of the angled design of the drill at the time the provisional patent application was filed and that its product always included this feature; however, failure to describe it in the provisional application meant that New Railroad did not get early priority on this part of the invention. Its provisional application was effectively worthless.

BENEFITS OF AN ATTORNEY-DRAFTED PROVISIONAL

Knowing how much detail and what types of detail to include in a provisional patent application is difficult for parties unfamiliar with

---

1. New Railhead Mfg., L.L.C. v. Vermeer Mfg. Co., 298 F.3d 1290 (Fed. Cir. 2002)

patents. In some cases, I have worked with clients who originally filed provisional applications that were a single paragraph abstract taking up less than a page—clearly not enough detail to provide any protection. On the other hand, I have worked with clients who originally filed massive provisional applications on their own with hundreds of pages of drawings, exhibits, and descriptions, yet still managed to leave out details that were important for patent protection. This occurs most often with computer-related technology because these inventions are protected by claiming a series of method steps that are performed by one or more devices. When describing how a piece of software works or how devices work together to achieve a goal, inventors often only describe such an invention in terms of user experience or capabilities, and not as a series of specific method steps that serve to provide such functionalities.

Working with a patent attorney therefore provides an assurance that a provisional patent application will not only describe an invention in sufficient detail, but also that it will describe the invention with the correct types of details that are relevant to patent protection. In addition to having the right level of detail, describing details of an invention in the wrong way can erode the scope of protection that a patent will provide. Because all patent documents eventually become public record, seemingly innocuous statements made in patent applications and during the examination process can be used against patent owners to prove noninfringement or even invalidity of issued patents during enforcement actions. Therefore, wording must be carefully chosen, even in a provisional patent application.

At the same time, not all companies or inventors can afford to have a top-notch attorney-drafted provisional patent application in the early development stages when patent protection is essential. This does not mean that patent protection should be forfeited or that a company should not move forward. Instead, the solution is to work closely with a patent attorney to get the best protection possible within the budget that is available while being cognizant of the risks that are inherent in lower-quality provisional patent applications. Luckily, there is a gradient of options available for essentially any budget.

OPTIONS FOR DRAFTING A PROVISIONAL PATENT APPLICATION

At the low end, inventors can draft their own provisional application and file it for only $65 or $130 in government filing fees.[1] On the high end, a top-quality provisional application can cost as much as a top-quality nonprovisional application that reaches into tens-of-thousands of dollars. While the do-it-yourself version is rarely a good option, the ultra-high-end version is also a poor choice for the vast majority of companies. The reality is that there is usually a balance in the middle where protection is maximized for a reasonable budget based on the type and complexity of the invention. This is where working with a carefully chosen patent attorney is important because you want to be sure that you are not sold on a cheap provisional application that provides little value or sold on an exorbitantly expensive application that wastes important working capital.

OPTION 1: DO IT YOURSELF COMPLETELY

Although not recommended, drafting and filing a provisional application without the assistance of a patent attorney may be necessary for inventors with extreme budget limitations. Fortunately, there are many good resources for drafting disclosures that can be filed as provisional patent applications, including the guidelines presented later in this chapter. The provisional filing process at the USPTO is outside of the scope of this book, but it is relatively easy to figure out, especially when using one of the many available self-filing guides.

However, as discussed in detail above, self-drafting and filing a provisional application comes with high risks, even for inventors who are highly experienced with technology or have some past experience with patents. The numerous pitfalls and quirks of patent law are not avoided because patent attorneys are smart, but because they have extensive experience with the complex and often counter-intuitive world of patents. Given that even basic assistance is available to inventors at relatively low cost, or even for free in some initial consultations, inventors

---

1. Provisional filing fees for small and micro-entities as of September 2014.

with budget limitations should still visit a patent attorney before settling on a self-drafted provisional patent application.

OPTION 2: ATTORNEY-COACHED PROVISIONAL

Some patent attorneys offer a service of coaching inventors in drafting their own provisional patent applications. For a flat fee or hourly rate, the attorney will teach inventors how to draft a provisional application and may even critique and edit one or more drafts before it is eventually filed by the attorney or inventors themselves. This method seems attractive because it allows the majority of the work to be done by the inventors, while providing some attorney oversight.

However, attorney coaching tends to be an inefficient use of time and money compared to other options. For example, when I worked at a firm that offered coached provisionals, I found that the outcome was rarely a positive one for the cost. The time it took to teach drafting lessons and then edit and comment on drafts often rivaled the time it would take for an attorney to draft an application. The coached version was always worse than a version that could have been drafted by the patent attorney at the same cost. The problem is that formally teaching someone to draft patent applications is time consuming for both the teacher and student, and any benefit gained by inventors taking on the bulk of the drafting work is strongly outweighed by the high cost compared to the strength of the provisional application that results.

Therefore, the best balance of cost and patent protection is found in two options. First, where inventors draft an application themselves with minimal attorney guidance and the document is filed by the attorney, possibly after some brief editing, but with the knowledge that patent protection may be thin. Alternatively, inventors provide a patent attorney with invention disclosure documents and the attorney drafts a provisional application on a set budget. These two best options are described below.

OPTION 3: FILING AN EMERGENCY PROVISIONAL APPLICATION

Where time and budget are extremely limited, a good option is to file what I call an "emergency" provisional application. For example, where inventors are making public disclosures within a few days or if there is no budget for an attorney-drafted provisional application, then a good choice is to first file papers drafted by inventors with the intention of drafting and filing a comprehensive attorney-drafted provisional immediately thereafter or once suitable financing has been established.

For example it is not uncommon for new clients to meet with me days before their business goes live on Kickstarter, before a paper on their work publishes, or before an important meeting with venture capital or angel investors. In these cases, where time is short, an emergency provisional can be filed that is specifically tailored to afford protection against loss of patent rights from these specific public disclosures. A provisional application is drafted that has the same content as these public disclosures, along with any other materials that the company has readily available. These documents are reviewed to remove any statements that could be damaging to patent rights and then filed as is. For presentations, the slides and a speech script are filed with little or no modification. For scientific or engineering publications, an exact copy of the paper that is being published can be filed as a provisional patent application.

However, this is done with the knowledge that this provisional application only provides limited protection that is useful only for a limited time. Ideally, work on a comprehensive attorney-drafted application should begin immediately after the emergency provisional is filed. However, for companies with extremely limited budgets, the goal is to secure financing from the public disclosure being made so that a suitable budget for a comprehensive attorney-drafted provisional can be prepared as soon as possible.

OPTION 4: PATENT ATTORNEY DOES THE DRAFTING AND FILING

The best option for filing a provisional patent application is having a patent attorney draft one based on invention documents and an inventor interview. Fortunately, because of the minimal filing requirements of a provisional application, a suitable attorney-drafted option is typically well within the budget of just about any business. For each unique invention and company, there is an ideal budget range that provides the most value for the invention based on the company's needs. Sometimes a large patent budget is optimal, whereas sometimes a smaller investment is optimal. The key is to work with a trusted patent attorney to determine what the ideal budget should be for your unique invention and business goals.

Providing good invention disclosure materials is one way to increase the value and minimize the cost of both provisional and non-provisional applications and in some cases such documents may be directly filed as an emergency provisional application in a pinch. With the requirements for a provisional application in mind, along with the most efficient ways of drafting one described in the first half of this chapter, the following provides a guide for generating documents that will be useful in the patent-drafting process.

TIPS FOR DRAFTING AN INVENTION DISCLOSURE

The primary purpose of an invention disclosure is to efficiently provide your patent attorney information about an invention that can be used when drafting a provisional or nonprovisional patent application. Nonprovisional applications include additional filing requirements as discussed in coming chapters, but an invention disclosure for both provisional and nonprovisional applications is essentially the same because both applications require a description that allows one of ordinary skill in the art to make and use the invention.

Although drafting formal invention disclosure materials is helpful for efficiently conveying information to your patent attorney, be mindful that drafting formal documents of any kind is completely optional. In fact, complete patent applications for many inventions can be drafted from nothing more than a short informal inventor interview. Documents can also be extremely rudimentary and, as mentioned in previous chapters, I have even drafted patent applications based on nothing more than some scribbles on a cocktail napkin. Patent attorneys are experts at taking whatever information or documents available and converting that into a patent application that best protects the invention at hand.

There is no standard way to disclose inventions to a patent attorney, and inventors should not be intimidated or concerned that they are somehow providing documents that are incorrect or not formal enough. Drafting invention disclosure documents should be an enjoyable experience that happens relatively quickly and, if not, you are likely over-thinking the process. Moreover, drafting invention disclosure documents should not be time-consuming for the vast majority of inventions.1 Anything more than a few days or weeks of delay in the filing of a patent application is unwise.

When putting together invention disclosure materials, it is important to keep in mind that they are intended to allow one of ordinary skill in the art to make and use the invention. In other words, you should be able to give this document to average people who work in the field of the invention and they should be able to make and use the invention without undue experimentation or additional guidance.

DRAWINGS, FIGURES, AND IMAGES

Because patent applications primarily include drawings and a description in reference to the drawings, it helps to follow this format when drafting invention disclosure documents. If possible, drawings should be black-and-white line drawings (no color or grayscale) so your attorney has the option of using these drawings directly in a patent application. If drawings are not completely black and white, they can be used, but must be converted to black and white, which can incur significant additional drafting cost and delay. Computer-aided drawing (CAD) programs allow for black–and-white line drawings, as well as software for drafting flowcharts and graphs. Computer drawings are not necessary, however, as even simple images made with black pen on white paper would be just fine. In the end, it is advisable to work with the medium that provides the most comfort. Do not be intimidated if your artistic skills are less advanced than a third-grader. In the end, the goal is to convey ideas—polished drawings are not necessary.

---

1. Chemical and biological inventions tend to require larger and more formal invention disclosures; however, developing these inventions over time typically includes documentation in lab notebooks that can be easily converted into an invention disclosure. The specifics of invention disclosures for chemical and biological inventions are discussed later in this chapter.

If you have ever looked at patent drawings, you probably noticed that the drawings have a set of numbered lines that label various parts of the figures, with corresponding references to these numbers in the description. Use labels in drawings if it comes naturally, but avoid being trapped by the style and form of patent drawings. It can be a fun challenge to emulate the style of patent drawings, but I find that inventors who do waste precious time and sacrifice content for form. Patent drawings should therefore only be used as inspiration and not as a template.

For mechanical devices, include images from all relevant perspectives, including sides, top, bottom, and the like. Include close-up images of important features if necessary. For products that move or assume multiple configurations, show the device in these different configurations and, if relevant, show what the device looks like transitioning from one configuration to another. For devices that interact with another object or a user, try to show this as well. If drawings are not possible or easy to make, just take pictures or make a video.

For software and other computer-related inventions, provide images that depict the user experience. For example, for a cell phone app, show the user interface and the important menus, screens, or other functionalities it might have. Screenshots or drawings of an existing prototype are great, but if the idea is at early development stages, rough sketches of an interface are fine. Those who are proficient with making advanced diagrams such as block or data-flow diagrams, should consider adding these as drawings, if possible. Show the software logic or algorithms performed and illustrate how data flows between multiple devices, where applicable.

Again, only prepare drawings, figures, and images if they come easily. This is not the time to learn a new CAD program or buy a book on drafting software diagrams. If you are not proficient with making drawings on a computer, make sketches by hand, even if they are crude. If hand drawings are difficult and unwieldy, pictures or videos work too. However, if preparation of any of these materials does not seem relevant to your invention or will be overly difficult or time-consuming, simply skip visuals altogether.

WRITTEN DESCRIPTION

Just as with drawing or figures, the written description should prioritize content over form. The primary purpose of an invention disclosure is to efficiently convey information. Although there are many ways to describe an invention, there is certainly no standard best way to do it. In fact, I am often troubled when inventors ask me for templates that they can use for preparing invention disclosure materials because I find that templates tend to act as a constraint more often than they provide useful guidance in describing an invention.

Using issued patents or published patent applications as a template is one of the worst ideas possible. Although patent applications are supposedly technology disclosures, they are in fact legal documents written in an archaic and formulaic style that has evolved over hundreds of years to maximize protection of inventions but not to maximize understanding of how an invention is made and used. This is a counter-intuitive and unfortunate reality of the current patent system, especially given that its purpose is to promote the useful arts and sciences by providing limited monopolies to inventors that are willing to show the public how to make and use their inventions. A thoughtful critique of the patent system and how it could benefit from change is outside the scope of this book, but the takeaway message is that you should not seek to copy the style of patents when writing an invention disclosure—let your patent attorney translate your disclosure into that style. Your goal should be to describe your invention in a way that is intuitive and easy for you. The remainder of the chapter provides guidance on efficiently drafting a comprehensive written disclosure that adequately and broadly describes an invention.

MINDSET FOR WRITING ABOUT AN INVENTION

When drafting an invention disclosure, there are some tricks to conceptualizing your invention that will significantly improve the quality of your disclosure and will come in handy during the patent process and in forming an understanding of how your invention is protected. For example, if you have an idea for a commercial product, you likely have at least a general concept of one or maybe a few variations of how the product would look and operate. Instead of thinking of these specific products as "the invention," consider them to be a small subset of a much larger invention, or as example embodiments of a much larger invention. This is illustrated in the Venn diagram in Figure 9.2, which depicts the scope of "the invention" as a large circle, with two example embodiments shown as smaller circles that are within the scope of "the invention."

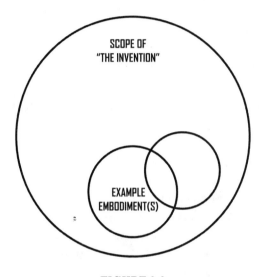

FIGURE 9.2

In fact, as discussed in greater detail in the coming chapters, this is how patents are intended to protect products. The protected product(s) should ideally be only a small portion of what the patent actually carves out as intellectual property owned by the applicant. This makes sense when considering that patents are filed to cover not only exact copies of a product being sold by a company, but also as many variations of the product as possible. Without this broader coverage, any competitor would easily be able to circumvent the patent, and not infringe it, by simply making minor nonsubstantive changes to the design or functionality of the product or by using different materials.

Therefore, when describing what you have invented, you never want to actually talk about "the invention," but instead want to discuss some specific example embodiments or variations of a larger and more nebulous invention. By doing this, you serve to expand the scope of protection that will be afforded in a subsequent patent application, including embodiments that have not yet been conceived or that may not even be technologically possible today.

Along these same lines, patents allow you to discuss any number of variations that you might contemplate for the product that you intend to protect, and you should endeavor to consider and include as many variations as practically possible. For most inventors, this tends to be the opposite of what occurs during product development. Instead of dreaming of all the possibilities, variations, and versions of a product that could be made in the near and far term, developing a commercial product often focuses on limiting a product to a single and specific embodiment that is constrained by numerous practical limitations. However, when seeking patent protection, it pays to put on the hat of an inventive daydreamer who is uninhibited by real-world limitations like manufacturing capabilities, production costs, development budget, product price-point, and the like.

Keep in mind that a patent term runs 20 years from the nonprovisional filing date, and that the technology landscape will see major changes during that time period. Although the future is hard to predict, attempting to account for capabilities and freedoms that future technology will offer can add substantial value to a patent application. For example, many developers of computer hardware and software struggle with the size, processing speed, and battery life limitations of current technology. History has shown, however, that devices have become consistently smaller and faster over time, in addition to acquiring a longer battery life. Projections indicate that this trend will continue for some time to come, with the theoretical limits of device size and processing power being far in the future. Given that most types of technology progress exponentially instead of linearly, inventors should plan for a world where the devices that implement their invention or run their software will be tiny, inexpensive supercomputers compared to the devices available today. By inadvertently shackling an invention with the limitations of today's technology, inventors all but guarantee that their patents will become obsolete during their term.

DETAILS TO INCLUDE IN AN INVENTION DESCRIPTION

We have established that the description of an invention must be sufficient to enable one of ordinary skill in the relevant art to make and use
the invention, but the types of details that are required to satisfy this
requirement are likely far from clear. Generally, the details required for
a patent application and therefore the details that need to be disclosed
to a patent attorney fall into three categories of inventions: mechanical,
computer, and scientific.

For mechanical and computer-related inventions, the level of detail
can be surprisingly low when it comes to an invention disclosure, and
inventors should not be intimidated even if they do not consider themselves to be proficient in a given technology field. This is especially
true when working with a patent attorney who has experience in the
field, because he or she can fill in the gaps where an invention disclosure might be lacking. Accordingly, when describing an invention, it
is completely acceptable to indicate that you are not sure how a certain
part of an invention will be implemented. In some cases, it may simply
be that there is a choice between several options and you are unsure
which option to choose. Just describe the different options and indicate
that any one is possible. In other cases, the inventor may have no idea
how a given aspect of an invention would be implemented. Sometimes,
there is no need to provide additional explanation because it would be
understood by one of ordinary skill in the art and the patent attorney
can fill in some examples. Other times, it may be necessary to work out
such details before the invention is ready to be patented.

For mechanical devices, it is typically sufficient to provide details
on how the device is configured and used. For example, refer to images
or drawings of the device from different perspectives and describe all
pieces of the product and how the pieces are interconnected. Provide a
step-by-step description of how the product is used. Where the product is operable to change configurations, describe in detail how the
change occurs in terms of how various parts move and what causes
them to move.

On the other hand, details such as specific dimensions are not
necessary unless specific sizes or proportions are a novel part of the

invention. Similarly, specifying exactly what materials are used is often unimportant unless the choice of specific materials is the novel part of the invention. However, a general description of possible materials should be provided if possible.

For computer-related inventions, a good place to start is with the devices involved. Is there only a single device or are there multiple devices? How are these devices connected and how do they communicate? Via a WiFi network, the Internet, or a cable? Also, do these devices have any special capabilities and must they have specific hardware? Must they have a screen, a speaker, or GPS capabilities? Can the device be off-the-shelf or must it be custom?

Next, walk through the user experience from beginning-to-end and describe some examples of where and how the invention can be implemented. If possible, describe communications that occur between devices and how data is processed and stored during each part of the user experience regardless of whether or not it is transparent to the user. Describe the software logic or algorithms that support important functionalities. In some cases, these details may not have been fleshed out or the inventor may have no clue about how these things are done. Again, there is nothing wrong with not having all the answers here. Simply describe as much as possible and move on.

On the other hand, details such as specific code language that implements functionalities or algorithms are rarely relevant. There is no need to have written code that embodies the invention, nor is it necessary to provide specific code in the vast majority of patent applications. Similarly, specific hardware choices are rarely important. There is no need to mention that a system runs on Windows or has a certain model of processor or a specific amount of memory. A higher level discussion of functionalities and device capabilities is sufficient, and one of ordinary skill in the art is presumed to know how to find or build a device that will implement the described functionalities or have the needed capabilities. These types of details can be left out unless they are extremely important to the novelty of the invention.

For chemical, pharmaceutical, and biological technology, much more detail is typically required. These invention disclosures should have content that is similar to a scientific paper, where the methods and materials used to setup example experiments are detailed along with the results that are gathered. Since these inventions typically involve compounds or biological materials that have certain properties or can affect a patient in certain ways, some semblance of proof is required that the invention works properly. Luckily, development of these inventions often inherently includes such documentation, and so the content of laboratory notes, reports, and papers is usually sufficient as an invention disclosure.

BACKGROUND SECTION

An optional, yet beneficial part of an invention disclosure is a background section. Nonprovisional patent applications are required to have this section and provisional patent applications often include one as well. Taking the time to draft an invention background is worthwhile for several reasons. First, as discussed in coming chapters, the background is the one section of a patent application where inventors can save time and money by drafting the content themselves and inventor-drafted backgrounds can often be directly pasted into a patent application with only a bit of editing. Given that less attorney drafting time translates directly into lower cost, writing a background section is a great way to save money without sacrificing patent quality.

Additionally, the process of writing a background section is a helpful exercise for inventors and growing companies, because it requires a clear articulation of what the current and historical technology landscape is and why it is deficient. Where specific prior art is known or was discovered during a search, it provides an opportunity to analyze these references. Doing an analysis of why and how an invention serves unmet needs in the current marketplace is a key analysis that should be done by any growing company, and writing a patent background section provides an opportunity to do this.

Plus, the background is the first main section in patent publications and is therefore often the only part of a patent that people will completely read through. The background therefore provides a unique opportunity to speak to potential customers, competitors, investors and business partners, so marketing messages that tout a product or business are well placed in the patent background.

The first rule for drafting a background section that can be included in a patent application is to never discuss the invention. The background is not a summary, abstract, or opportunity to compare and contrast the invention to what is already out there. It only provides a background on the invention's field of art. A good approach is to start with a general discussion of the field of art, which might include some history of the field and an identification of the problems solved by existing products in the field. Then provide some examples of products or inventions and indicate why these products are deficient. The final sentence should simply state that additional inventions are needed to fulfill unmet needs. The background need not be more than two or three paragraphs.

For example, if the invention was a new bicycle braking system, the first paragraph could be an introduction that mentions that bicycles were first introduced in the early 1800s in Europe and that there are now a billion bicycles in the world, with nearly half being in China. Given that the innovation is related to brakes, the next paragraph could introduce disc and caliper brakes as being the most common types of braking systems, followed by a brief description of at least one flaw these brakes have. If there are specific patents or products that are similar, another paragraph can briefly describe these inventions and briefly indicate why they are deficient. The final sentence would simply state that, based on these indicated deficiencies, there is a need in the art for improved braking systems.

For additional examples of the level of detail required in various types of invention disclosures along with examples of ideal drawings that should be included, visit www.PatentsDemystified.com.

CHAPTER 9 SUMMARY

- Patent applications must describe an invention with sufficient detail such that one of "ordinary skill in the art" could make and use the invention.
- The USPTO will give a filing date to any document filed as a provisional patent application, but a poorly drafted document will give no protection and is effectively worthless.
- Provisional patent applications can include informal documents, but applicants only get priority for the details that are adequately described, so the content of a provisional application is just as important as a nonprovisional application.
- Applicant-drafted invention disclosure documents can be filed as a provisional application in an emergency or where there are severe budget constraints, but an attorney-drafted provisional application is best for maximizing protection.
- Invention disclosure materials are often optional or can be much simpler than expected, so applicants should not delay meeting with a patent attorney if preparing such documents seems daunting or overly burdensome.
- Content is more important than form in invention disclosure documents—any format is fine as long as it conveys how an invention is made and used.

10

Working with a Patent Attorney to Draft a Provisional Patent Application

"However, patent lawyers who work with small clients have said that they advise their clients not to treat a provisional application any less seriously than a full patent application. If there is part of an invention that is left out of the provisional application, that will not be protected."

—Senator Dianne Feinstein during debate on the America Invents Act, March 2, 2011

"They tell me that the balm of 'cheap provisionals' is snake oil, because a provisional still has to meet certain legal standards. . ."

—Senator Harry Reid during debate on the America Invents Act, March 2, 2011

As discussed in Chapter 9, working with a patent attorney to draft a provisional patent application gives the best balance in terms of cost and patent protection. The unique business needs of a company and complexity of the invention will dictate an ideal budget range, but it is still often possible to draft and file a provisional patent application for less than this ideal budget, if necessary, without disproportionately sacrificing quality.

These budgets, however, are estimated based on an expectation of a "typical" drafting process between the inventors and attorney. For projects that are billed hourly, an expected budget can be exceeded if unexpected drafting and editing time is necessary. Flat-rate projects also suffer because where additional time is required elsewhere in the project, other important aspects of the drafting and filing process may receive insufficient attention. On the other hand, if the process moves faster than expected, extra time can be used to make a higher quality application or may simply result in the project costing less. Accordingly, using time in the most efficient way possible is beneficial, regardless of whether the billing model is fixed, hourly, or a combination of the two. The result is a higher quality patent application that often comes at a lower-than-expected cost. Getting the most value out of a patent budget therefore requires an understanding of what a "typical" provisional patent drafting process is and then learning some insider tips on further streamlining the process.

TYPICAL DRAFTING PROCESS

When providing a cost estimate or setting a flat rate, patent attorneys assume that the provisional drafting process will have certain parameters. Although there is no standardized method for drafting a provisional patent application, the process tends to be the same, and some methods are more cost-efficient than others. Unfortunately, patent attorneys do not always clearly articulate what their assumptions are, and clients often have different expectations of how the drafting process will progress. The following sets out the most common and most efficient steps to drafting a provisional patent application.

INVENTION DISCLOSURE DOCUMENTS AND INTERVIEWS

As discussed in previous chapters, the provisional application drafting process begins with an invention disclosure, which may include providing invention disclosure documents and a face-to-face invention disclosure session with the patent attorney. Invention disclosure documents are not always required, but providing an invention disclosure, as described in Chapter 9, helps ensure that the attorney has a complete understanding of the metes and bounds of the invention.

An invention disclosure interview can replace the need to provide invention disclosure documents, and can often be more efficient when the inventors are unsure of what documents to provide or are having a hard time drafting documents. For example, if you find the process of sketching and describing the invention daunting, time-consuming, and unenjoyable, it likely makes sense to skip the invention disclosure documents and instead make a disclosure in an interview.

However, the ideal process begins with providing invention disclosure documents to your patent attorney in a timely manner, and then following up with an interview. An initial set of disclosure documents allows the attorney time to fully digest the invention and then provides an opportunity to ask additional questions. Even with the most comprehensive invention disclosure materials, I always learn new and important aspects of the invention that were inadvertently left out of the disclosure or were initially unclear. Accordingly, while provisional patent applications can be drafted based on documents or an interview alone, the application will almost certainly be higher-quality with a combination of both.

THE APPLICATION DRAFTING PROCESS

Once the substance of the invention has been conveyed, the attorney then begins to draft the patent application, which often takes one to four weeks. Drawings and other figures typically provide a foundation for the application and so are often drafted first. For physical inventions, or software inventions that include a user interface, the quality of the application can be significantly improved and cost can be reduced if the right drawings are provided. The key is to provide drawings or other images that the attorney can easily use and convert into patent drawings. However, providing drawings in the wrong format can actually incur more cost.

The cardinal rule of patent drawings is that they must be black and white, with no color or grayscale. For inventors that use CAD programs to sketch or prototype their invention, this will be an easy task because these programs are made to output black-and-white drawings in various perspectives that will be relevant to a patent application. Ideally, isometric views and top, side, and bottom views should be provided along with any other view available that is relevant to understanding the invention. Err on the side of providing too many images so that the attorney can select and include the most relevant ones.

If you do not already have CAD drawings of the invention, do not take the time to draft them from scratch unless you are proficient with the program and are able to generate drawings quickly. Instead, provide black-and-white hand-drawings or even pictures of a prototype at various relevant angles, which the attorney can convert to line drawings with the assistance of a patent draftsperson. Drawings should ideally be provided in a conventional digital format, but hard copy also works. Individual attorneys may have specific format preferences or capabilities, so be sure to ask your attorney if you are unsure of what medium to use.

Providing images of a computer interface can be tricky but helps to increase the quality of a patent application. The main issue is removing all color and grayscale from such images, because most user interfaces are designed to be rich and colorful. Do your best, and provide screenshots or layouts of the important screens that show how a software product is used.

If the idea of providing drawings or images has you concerned, rest assured that you can get by without providing a single sketch or image of the invention. Skilled patent attorneys will be able to work with whatever you have (or don't have) and still draft a great application. Providing a comprehensive set of black and white line drawings is only the "ideal" case. It is unnecessary to delay the patent process or not move forward simply to craft an ideal set of drawings.

Looking at the drawings from issued patents or engineering diagrams can be a good way to get inspiration for what kinds of drawings might be useful, but do not aim to completely replicate patent drawings when providing documents. For example, such drawings include figure numbers and numbered lines that point to specific parts. These annotations are not useful when crafting patent drawings and will actually need to be removed by the attorney before new figure numbers and labels can be added. Accordingly, even if labeled drawings are provided in an invention disclosure, clean drawings should also be provided.

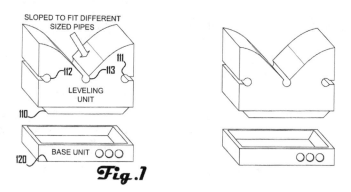

FIGURE 10.1   Examples of annotated and clean drawings

After generating suitable patent drawings that include figure numbers and part labels, the patent attorney then drafts the complete patent specification, which includes a background, abstract, a brief description of the drawings, and a detailed description that references the drawings. Provisional applications have no required format, but usually follow this same format so that they can be more easily converted into a nonprovisional application down the road. Other sections that might be present in a provisional patent application include a brief description of the field of the invention, a summary of the invention, and patent claims. Once the drawings and patent specification are fully complete and edited, a polished draft is provided for applicant review and approval before filing.

REVIEW OF THE PROVISIONAL PATENT APPLICATION DRAFT

Reviewing the draft of the application is where a patent project most often goes over budget. The review process sometimes takes longer than expected, but it is more often the case that this additional cost is unnecessary and completely avoidable.

For example, when estimating the time it will take to finalize and file a patent application, only an hour or less is allocated to this phase of the project. In other words, once a draft of the patent application is delivered for review, 95% or more of the time budget will have already been expended. Accordingly, an efficient review of the application is imperative to stay within budget.

Before going into the mechanics of cost-effectively reviewing a draft of a patent application, it is useful to step back and conceptualize the patent drafting process. With such a framework in place, it is easier to not only understand how to make the process more efficient, but also *why* such a method is more efficient. First, think of working with a patent attorney to draft a patent application like translating into another language. In other words, the job of a patent attorney is to translate a disclosed invention into a patent application that is written in the unique style and language of patent law.

What makes this counter-intuitive is that, on their surface, patent applications appear to simply be technical documents that describe an invention. In contrast, patent documents are legal documents that are written in a way designed to maximize protection of an invention while also guarding against invalidation and other attacks during litigation. Describing the technology in a way that makes sense is completely subservient to this primary purpose as a legal document.

Accordingly, drafts of patent documents should be reviewed for content and not for style. The focus of the review therefore should be to determine if the invention is disclosed correctly and as broadly as possible. Spelling and grammar should also be corrected where necessary, but direct editing of word choice, sentence structure, and other stylistic choices is not recommended. At best, stylistic edits make the document only subjectively better while still incurring additional cost because all changes are then reviewed by the patent attorney. At worst, such changes, even subtle ones, can negatively affect the scope of protection and create holes that can be exploited during litigation. Keep in mind that specific words, phrases, and sentence structures are often carefully chosen for legal reasons that might not be readily apparent.

For this reason it is advisable to not directly edit or add to patent document drafts, but to instead provide comments, questions, or suggested edits outside of the document in general form or citing to specific paragraphs. Using the translation analogy, it is more cost-effective to have the attorney consider comments and then translate them into patent-speak rather than reviewing edits made directly to a document, and then determining how to translate these edits into appropriate patent language.

In addition to confirming that the invention is accurately described in a draft patent application, it is also important to determine that the invention is described as broadly as possible. As discussed in previous chapters, patent applications can be drafted to describe numerous variations of an invention, and the protection afforded by a patent is increased by describing as many different variations as possible. Patent attorneys will always try to describe the subject invention broadly when drafting an application, but may inadvertently miss important variations or impose unnecessary limitations on the invention that inventors would recognize when reviewing a draft.

However, given that the vast majority of the drafting budget is assumed to have already been expended at the draft review stage, the budget may be exceeded if new embodiments or new invention details are introduced at the last minute that require a substantial amount of additional drafting time. This highlights the importance of making a full invention disclosure before application drafting begins and providing any updated material as soon as possible instead of waiting until the first draft is completed.

Once a final draft has been approved, the application can be finalized and the remaining filing documents can be prepared for filing.

FINAL STEPS FOR FILING THE APPLICATION AT THE USPTO

Once the provisional patent application draft has been reviewed, finalized and approved, a few final determinations need to be made before the application can be filed. The first is identifying the inventors and the second is formalizing ownership of the patent application.

INVENTORSHIP

Patent applications must name the inventors of the invention that is the subject of the patent application, and for provisional applications, this determination is made based on the invention described in the description and drawings. It is imperative to name all true inventors, and failing to name an inventor is just as detrimental as naming someone who is not an inventor. Both can result in invalidation of a patent if incorrect inventorship is asserted with deceptive intent. Plus, failing to add a true inventor can ultimately result in losing control of the patent rights.

For example, in 1985, Dr. InBae Yoon was granted U.S. Patent 4,535,773—he was named as the sole inventor. The patent related trocars, which are specialized surgical tools for making small incisions in a patient's abdomen during endoscopic surgery. Dr. Yoon exclusively licensed the patent to Ethicon, Inc., a medical supply company, which then sued its competitor, U.S. Surgical Corp., for infringement of the patent. However, U.S. Surgical discovered that Dr. Yoon had collaborated with Young Jae Choi, a resourceful electronics technician who had some experience with physics, chemistry, and electrical engineering, but no college degree. Among other inventions, Choi helped Dr. Yoon develop the surgical trocars that were the subject of the '773 patent, but their collaboration eventually soured and dissolved because Choi felt that Dr. Yoon was dissatisfied with Choi's contributions.

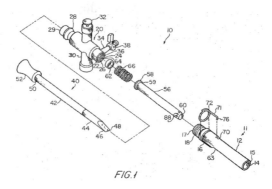

FIG.1

FIGURE 10.2   Drawing of a Trocar from U.S Patent 4,535,773

During the litigation, defendant U.S. Surgical recognized that Choi should have been named as a co-inventor and quickly solidified a retroactive license agreement with Choi for rights to make and sell the patented trocars. With Choi's cooperation, U.S. Surgical ultimately succeeded in amending the patent to name Choi as a co-inventor, and the Ethicon lawsuit was dismissed because of U.S. Surgical's valid license to use and sell the technology.[1]

This crafty maneuvering by U.S. Surgical comes from the default rules that give each of the named inventors an undivided interest in the entire patent, regardless of how small their inventive contribution is.[2] Moreover, each inventor is entitled to use the patent or license it to others without the consent of the other inventor(s) unless there is an agreement to the contrary.[3] Accordingly, in addition to correctly identifying who the true inventors of an invention are, it is also important to formally determine how patent rights will be controlled and owned—at least when the first patent application is filed, and ideally much sooner.

ASSIGNMENT

Inventors are the default owners of their inventions, even if they are employees of a company, working as a contractor, or are otherwise subservient to other collaborators. Inventors must therefore explicitly relinquish their rights in an invention. As discussed in detail in Chapter 4, this is often done with an assignment clause in an employment agreement or other collaboration contract that inventors sign before work ever begins. However, where the assignment of an invention is not done earlier, it should be done at the time of patent filing at the very latest. Assigning patent rights to a company (such as a corporation or LLC) gives the company control of patent rights instead of the unde-

---

1. Ethicon, Inc. v. United States Surgical Corp., 135 F.3d 1456, Ct. App., Fed. Cir. 1998.

2. *See, e.g., Ethicon* 135 F. 3d at 1465. ("Indeed, in the context of joint inventorship, each co-inventor presumptively owns a pro rata undivided interest in the entire patent, no matter what their respective contributions. Several provisions of the Patent Act combine to dictate this rule.")

3. 35 U.S.C. § 262. ("In the absence of any agreement to the contrary, each of the joint owners of a patent may make, use, offer to sell, or sell the patented invention within the United States, or import the patented invention into the United States, without the consent of and without accounting to the other owners.")

sirable default ownership rules that give inventors complete control. In some cases, inventors may be owners of a company and still retain control over patent assets, but in other cases some or all of the inventors may lose control of patent rights completely.

Even where assignment has occurred previously as part of another agreement, the inventors need to execute a formal assignment agreement specific to the invention of the patent application, which will be officially recorded at the USPTO to allow the assignee to be named as the patent applicant and owner of the patent in all patent documents. With the inventors of the subject matter of a provisional application identified and with a freshly executed patent assignment agreement, the provisional patent application can then be filed.

WHAT TO DO AFTER PROVISIONAL IS FILED

A finalized provisional patent application and its related documents can be filed at the USPTO quickly, and often times that same day. Since patent filings are predominantly done electronically, the applicants immediately establish a filing date and get an application number in an electronic filing receipt that is generated when the online filing at the USPTO is complete. It is standard practice to send this electronic filing receipt to the applicants, along with a copy of the application as filed, to officially confirm that the provisional application was received at the USPTO. Applicants should wait to receive a copy of this filing receipt from their patent attorney before taking any actions such as making public disclosures, public uses, or offers for sale.

"PATENT PENDING" STATUS
As of the day that the first patent application is filed (which includes provisional applications) the invention is considered "patent pending" and a company has the option of marking its products, publications or marketing materials with this designation. There are no strict requirements for this marking, but the most common markings are simply "Patent Pending" or "Pat. Pend." Using a "patent pending" designation does not improve patent rights or the ability to receive damages from potential infringers; it merely puts the public on notice that a patent

application has been filed in association with the product or invention and that copiers or unauthorized users may be liable for infringement if a patent later issues.

Using a "patent pending" designation is almost always beneficial, and as discussed in previous chapters, it can be an important marketing opportunity. For example, many companies have a standard practice of issuing a press release when patent applications are filed. In addition to providing increased exposure to consumers, competitors, and other stakeholders, such press releases can often boost the stock price of some publicly traded companies.

PATENT APPLICATIONS VS. PATENTS

When describing patent assets, be sure to describe their status correctly. Failure to do so is against the law and can result in serious consequences. For example, be careful not describe filing a patent application as "filing a patent". One files a patent *application*, but does not file a patent. Filed nonprovisional patent applications can mature into issued patents, but only after being examined and allowed by an examiner. Also, products can only be called "patented" once a patent application issues as a patent and that patent covers the product. Similarly, "patent pending" applies only when at least one patent application has been filed and remains pending. When in doubt, consult your patent attorney.

However, publicizing provisional patent filings should be done carefully so as to maximize the impact and benefit. Since "patent pending" refers to both provisional and nonprovisional patent applications as well as to design patent applications, it is often desirable to leave some ambiguity as to the type of patent application. Provisional or design patent applications can sometimes be perceived as inferior to or weaker than nonprovisional utility applications, so simply stating that a product or invention is "patent pending" avoids any potentially negative connotations.

Similarly, it is typically beneficial to not provide an application number and to instead only claim "patent pending" status. For example, provisional patent applications are not available to the public while they are pending and even nonprovisional applications are not publicly available until the application publishes. Providing a patent application number that does not lead to publicly available documents can be confusing, and at worst, might even lead some to believe that a company has lied about filing for a patent or that it accidently provided an incorrect number. For these and other reasons, and as discussed in more detail in coming chapters, it is best to only provide numbers for patent publications or issued patents.

Another marketing trick is to selectively claim that an invention or product is "patent pending" in both the United States and internationally after filing a provisional patent application. Such a statement is justified because provisional patent applications can be claimed as priority if foreign patent applications are subsequently filed. Provisionals therefore preserve the option to file for nonprovisional patents in foreign countries, just as they preserve the option to subsequently file a nonprovisional patent application in the United States. Such a statement can make it appear as though a company has a sophisticated international patent portfolio in the works.

However, such a designation should be used with extreme care and with patent attorney supervision. This statement becomes untrue once the provisional patent application expires and the opportunity to file foreign patent applications is lost. Claiming international "patent pending" status is then fraudulent. Plus, this strategy can backfire if people catch on that no foreign patent applications are ever filed after the provisional application expires.

ONE-YEAR DEADLINE

As previously discussed, provisional patent applications are merely placeholders and will expire one year from when they are filed. If a nonprovisional patent application is not filed within this one-year pendency, the ability to claim priority to this early filing date is lost. Provisional applications are not extendable, and there is no way to make the term of a provisional application last longer than a year. Re-filing the provisional application is possible, but again, the earlier priority date is still lost, which in many cases would result in loss of patent rights because of public uses, disclosures, and offers-for-sale that that occurred before or after the first provisional was filed.

Accordingly, when a provisional application is filed, the applicants should keep in mind that a clock immediately begins to count down to when the nonprovisional patent application must be filed. The cost of the nonprovisional application depends on how much the provisional application needs to be updated and converted to a nonprovisional application, and applicants should plan for this expense in the coming year. Similarly, the deadline for filing foreign patent applications is also one year from when the first provisional application was filed, so applicants need to plan for this expense as well if foreign patent protection is desirable.

Although the provisional application lasts exactly one year, work on a follow-up nonprovisional application and any foreign applications should begin as early as one to two months before this one-year anniversary. With only 10 to 11 months before further patent budget becomes necessary, applicants should keep this timeline in mind when it comes to product and business development. For example, where a company wants to market-test a product or secure investors before investing time and money in the nonprovisional patent application, the company should be prepared to complete the process within 10 to 11 months of filing a provisional application.

Additionally, as discussed in previous chapters, a provisional patent application only protects what is described in the application documents, and significant changes to an invention after filing a first provisional patent application may require a second provisional application filing before the one-year expiration of the first provisional patent application. For example, products are typically kept secret until a provisional application is filed, and afterwards, the scrutiny of customers, product developers, and other contributors can result in the originally conceived invention changing radically or in the conception of new innovations that stem from the original idea. A good provisional application will attempt to preemptively cover such changes or further developments, but sometimes the changes or additions are so great that the first provisional application no longer covers the new invention.

Companies should therefore check-in with their patent attorney during the year in which a provisional application is pending to determine if updates or a completely new application is warranted. In many cases, updating a provisional application can be relatively inexpensive and simply requires adding drawings or description to the existing document and filing the updated application with the USPTO. This second document can then be converted into a nonprovisional patent application at the one-year anniversary of the first provisional application, with priority being claimed to both provisional applications. Theoretically, several provisional applications could be filed within the one-year term of the first provisional application, but this is rarely necessary. The following chapter discusses the process of drafting and filing nonprovisional patent applications, regardless of whether a provisional application is filed first.

CHAPTER 10 SUMMARY

- The process of an attorney-drafted provisional application begins with applicants providing invention disclosure documents and/or an invention disclosure meeting.
- Within approximately one to four weeks, the patent attorney then crafts a complete provisional application and provides it to the applicant for review.
- After implementing any edits or additions to the provisional application, an assignment agreement is typically executed by the named inventors and the provisional application is then filed electronically at the USPTO.
- Within the one-year term of the provisional application, any changes or additions to the invention can be protected by filing an updated provisional application or such updates can be included in a nonprovisional patent application that is filed before the provisional application expires.

Working with a Patent Attorney to Draft a Nonprovisional Application

"The patent laws 'promote the progress of Science and Useful Arts' by rewarding innovation with a temporary monopoly. U.S.C.A. Const., Art. I, § 8, cl. 8. The monopoly is a property right; and like any property right, its boundary should be clear. This clarity is essential to promote progress, because it enables efficient investment in innovation. A patent holder should know what he owns, and the public should know what he does not."

—Justice Kennedy writing for the court in Festo v. Shoketsu, 535 U.S. 722, 730-31 (2002)

As discussed in the previous chapters, one way to begin the patent process is by filing a provisional patent application and then filing a non-provisional patent application within one year. In some cases, however, it makes sense to skip the provisional application step and begin the patent process by initially filing a nonprovisional patent application. Either way, the process of drafting and filing a nonprovisional is essentially the same aside from being able to use content from a provisional application as a head start.

In the previous chapters that discussed provisional patent applications, the focus was on drafting a document that broadly teaches how to make and use an invention. Having an adequate description of the invention is equally important in a nonprovisional application, but because it is substantively examined by the USPTO, the nonprovisional includes an additional part that will be center-stage during the examination process—the patent claims.

EXAMPLE PATENT CLAIM FOR A BICYCLE FRAME

1. A bicycle assembly comprising:
a main frame comprising a seat tube, a head tube and an intermediate tube connecting the seat tube and the head tube;
a sub-frame configured to rotate with respect to the main frame;
a shock absorber having first and second opposing ends and a first eyelet connected to the main frame at the first end and a second eyelet at the second end, the shock absorber defining an axis between the first and second end;
an extension body comprising a first end and a second end, wherein the second eyelet at the second end of the shock absorber is positioned within the first end of the extension body and the second end of the extension body is connected to the sub-frame forming a rear pivot of the shock absorber; and
a fastener to secure the shock absorber within the first end of the extension body, the first end of the extension body configured to prevent rotation about the axis between the second eyelet of the shock absorber and the extension body.

Unlike provisional patent applications where essentially any format is allowed, nonprovisional patent applications have specific requirements. The format of an attorney-drafted provisional application discussed in the previous chapter is essentially the same as about two-thirds of a nonprovisional application. The benefit of drafting a provisional patent application with this same format is that when it comes time to file the nonprovisional application, the drawings and detailed description already meet the exacting requirements for a nonprovisional application. However, nonprovisional applications have an additional section at the end of the application called the claims. Individual patent claims are a single sentence and each one is sequentially numbered. An example of a patent claim for a bicycle is shown above.[1]

As you can see, patent claims are written in a unique style that is difficult to parse and understand. Luckily, learning to write in the strange language of patent claims is not necessary and should be left to your patent attorney. However, understanding how the patent claims function in the nonprovisional application is essential to making effective contributions to the patent process.

UNDERSTANDING PATENT CLAIMS

The glob of seeming incoherent text in the box below is the legal description for my house. More accurately, it is the legal description of the land that my house sits on. Believe it or not, this text precisely describes a roughly trapezoidal parcel of land in Seattle with three straight sides and one curved side that abuts public property where a bridge resides. In the unlikely event that I get into a land dispute with my neighbor, Jim, a land surveyor could come out and identify where my property ends and where his begins.

---

1. Claim 1 of U.S. patent 8,622,411

PAR E SEA SP #81217-0246 REC #8208040521 SD SP DAF POR OF SW
1/4 OF NW 1/4 SEC 19-25-04- BEG NW COR COLONIAL VIEW ADD TH S
00-05-49 W ALG W LN SD PLAT 263.47 FT TO NLY MGN N QUEEN ANNE
DR TH S 68-29-06 W ALG SD MGN 79.07 FT TO BEG OF CRV RGT HAV
RAD 1970 FT TH WLY ALG SD CRV 250.93 FT TH S 75-46-59 W 159.27
FT TO E MGN 2ND AVE N AS ESTAB BY ORD #76424 TH N 09-12-30 W
ALG SD MGN 283.30 FT TO WLY PROD OF S LN OF PAR DESC IN REC
#5419630 TH S 89-52-23 E ALG SD S LN 137.43 FT TO W LN OF PAR
DESC IN REC #5396991 TH S 00-06-38 W 26 FT TO SW COR SD PAR TH S
89-52-23E 80 FT TO SE COR SD PAR TH N 00-06-38 E ALG E LN SD PAR
& ALG PAR DESC IN REC #5081786 DIST OF 126 FT TO NE COR SD PAR
TH N 89-52-23 W ALG N LN SD PAR 24 FT TO SLY PROD OF WLY MGN OF
MAYFAIR AVE N TH N 00-06-38 E ALG SD LN 30 FT TO N LN OF SW 1/4
OF NW 1/4 SD SEC 19 TH S 89-52-23 E ALG SD N LN 54 FT TO NW COR OF
PAR DESC IN REC NO 4504314 TH S 00-06-38 W 53.50 FT TO SW COR SD
PAR TH S 89-52-23 E 80 FT TO SE COR SD PAR TH N 00-06-38 E 53.50
FT TO NE COR SD PAR TH S 89-52-23 E 184.72 FT TO POB

Patent claims are like a legal description, but for an invention. In other words, patent claims define the metes and bounds of the intellectual property of a patent, just like a legal description defines the metes and bounds of a parcel of land. More specifically, patent claims define the elements of an invention that a patent owner can exclude others from making, using, selling or offering to sell in the United States.

The term "intellectual property" is very appropriate because how patent rights operate is analogous to a real property like real estate, except that the property rights of a patent are intangible or intellectual. Previous chapters introduced the idea that all patents are not created equal, and that some patents are broad, whereas others are narrow. Using the analogy of real property, it is possible to better understand what this means.

For example, assume that your friend just bought a new empty parcel of land in the United States and wants you to guess how much it might be worth. There are two main questions that you would want to ask first before starting any analysis: (1) where exactly is the land located and (2) how big is it? Location is obviously extremely important—land in Manhattan is certainly more valuable than the desert in the middle-of-nowhere in Arizona. Size makes a big difference as well—just one acre in Manhattan is more valuable than 100,000 acres in Arizona.

The value of a patent is generally determined in the same way, but instead of physical location being a factor, the type of invention makes a difference. For example, how big is the market for the products covered by the invention, and how much do consumers pay for them? Naturally, patents on products that are expensive and used by many people will be inherently more valuable than patents on cheap products that are used by very few people. Although companies will certainly try to develop the market for their products, market value is not defined by a patent directly and will not be changed based on how a patent application is drafted or what happens during examination. This parameter is independent of the patent process.

On the other hand, how patent claims are drafted and the language of the claims that survives the examination process directly affects the amount and scope of exclusivity that the issued patent provides. In other words, just as you can own land that is small or large, patents can afford broad or narrow protection of a given invention, which is almost exclusively determined by the words in the patent claims. However, as discussed in the coming sections, the patent examination process is more like a Wild West land grab, where you can attempt to lay claim to large swaths of intellectual property landscape and might be able to keep some or all of it if you can successfully defend and justify your right to it.

FIGURE 11.1   Example of a Bicycle

Understanding what makes claims broad or narrow can be a bit counter-intuitive because less is more. For example, take a look at the three alternative patent claims shown below that could be used to claim the bicycle depicted above in Figure 11.1.

1. A vehicle comprising:
    a front and rear wheel;
    a frame including handlebars and at least two bars spanning between the wheels;
    a pedal assembly coupled with the rear wheel via a chain; and
    a pair of brake levers disposed on a set of handlebars.

2. A vehicle comprising:
    a front and rear wheel;
    a frame including handlebars and at least two bars spanning between the wheels; and
    a pedal assembly coupled with the rear wheel via a chain.

3. A vehicle comprising:
    a front and rear wheel;
    a frame including handlebars; and a pedal assembly coupled with the rear wheel via a chain.

Claim 3 is the broadest and, therefore, best claim because it has fewer elements. Claim 1 is the narrowest and thus the worst claim because it has the most elements. This makes sense considering the requirements for infringement of a patent claim, which is that the allegedly infringing device must satisfy all elements of the claim. The allegedly infringing device can have more elements than those recited in the claim, but if even one element is missing, then the claim is not infringed. One claim is considered better than another claim when it is easier or more likely to be infringed by competitors that try to copy a product. In other words, a claim is broader when it covers a wider variety of variations of the product being patented.

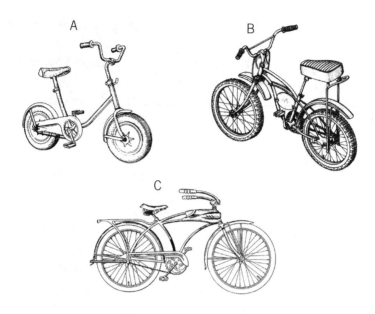

FIGURE 11.2   Three Potentially Infringing Bicycles

For example, suppose that the three bicycles shown in Figure 11.2 are competitor products that have recently come on the market. If these three products are compared to the three alternative versions of claims used to protect the bicycle shown above, it becomes clear that claim 3 is the strongest. Claim 1 has one element that is "a pair of brake levers disposed on a set of handlebars." All three of the competitor products shown here have handlebars, but none has brake levers on the handlebars, and so, for at least this reason, none of these competitor products will infringe claim 1. For claim 2, the bicycle shown in Figure 11.2A will not infringe because it does not have "at least two bars spanning between the wheels." On the other hand, because they have "at least two bars spanning between the wheels" along with all of the other elements, the bikes of Figures 11.2B and 11.2C will infringe claim 2. Specifically, the second and third bikes have "a front and rear wheel; a frame including handlebars and at least two bars spanning between the wheels; and a pedal assembly coupled with the rear wheel via a chain." Since all these elements are present, the second and third bike infringe claim 2.

On the other hand, all three bikes will infringe claim 3 because each of the bikes has "a front and rear wheel; a frame including handlebars; and a pedal assembly coupled with the rear wheel via a chain." Claim 3 is therefore the broadest claim because it has fewer elements and is more likely to be infringed.

So, if having fewer elements makes claims is better, why would patent applicants ever want to have anything more than the most basic of claims with the fewest elements possible? The answer lies in the examination process. For a patent application to be allowed by the Examiner, the claims must include a set of elements that is new and nonobvious in view of the prior art. Fewer claim elements make for broader claims, but broader claims are more likely to be rejected by the examiner. Accordingly, as discussed in detail in the next chapter, the ultimate goal of the examination process is to work with the examiner to get the application allowed by presenting patentable claims, but with the fewest elements possible, so that the patent gives a broader monopoly.

However, when thinking about patent claims, it is important to keep in mind that infringing patent claims are completely different from the analysis that a patent examiner does during examination of a nonprovisional application. Patent infringement occurs when a product has all of the elements of at least one claim of an issued and active patent. Alternatively, where a patent claim is for a method, infringement occurs when a process is performed that has all of the elements recited in the patent claim. In contrast, the examination process requires an examiner to agree that the pending patent claims recite a set of elements that are new and nonobvious over the prior art.

Given that the patent claims are the most important part of the nonprovisional application, the drawings and written description that together form the specification are merely there to support the patent claims. For example, the Venn diagram in Figure 11.3 illustrates the relationship of the specification and claims. The large circle represents the scope of what is described in the specification, and the smaller boxes represent the scope of some of the claims of the patent. Outside the circle is other technology that is not described in the specification.

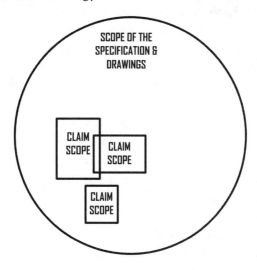

FIGURE 11.3   Relationship of the Specification and Claims

The important takeaway from this relationship is that the scope of the claims will always be narrower than the scope of the specification as a whole, and the scope of the claims must be encompassed within the scope of the specification. Recognizing this relationship between the specification and claims is one of the biggest insider secrets to understanding patents. For example, most people who read patents assume that what is shown and described in the drawings and specification defines what is protected by the patent and they overlook the claims because they are difficult to read. On the contrary, the protection afforded by the patent will *always* be substantially narrower than what is described in the specification as a whole. In other words, only a small part of what is described in the specification is actually covered by the claims, which are by definition the words that define the granted patent monopoly.

Recognizing this one fact about patents can be extremely valuable when it comes to licensing or enforcement of patents. For example, where companies are approached to license a patent, or are accused of patent infringement, they should know that the claims actually define what the patent and should not be intimidated just because the drawings and written description seem to describe technology that the allegedly infringing company uses. On the other hand, it is not uncommon for unsophisticated companies and those that are not represented by patent counsel to take a license or stop using certain technology simply based on what is in the drawings and description of a patent, when in reality the technology that they practice is not even close to being covered by the patent claims. In fact, I have been surprised by how many of my clients are approached by companies who insist on taking a license to my clients' patents when one would not actually be necessary. Instead of engaging a patent attorney to analyze the claims of the patent in question and determine whether a license would be needed to practice their technology, they enter into often expensive licenses that needlessly drain capital in favor of the patent holder. Moreover, I suspect that even more companies quietly choose to not practice certain technology or pursue a business idea because they incorrectly believe or fear that other companies own blocking patents.

Unfortunately, the convoluted language of patent claims and complex laws that govern patent infringement make it difficult for anyone but patent attorneys to effectively decode and translate patent claims into a picture of what would be infringement and what would not. Attempting to analyze patent claims alone is not recommended. Similarly, the process of formulating and writing patent claims is also a task that is best left to a patent attorney during the patent drafting process. However, applicants should be able to generally understand the claims of their patent application so that they can effectively contribute to crafting an appropriate overall patent strategy and drafting the best patent application possible.

DRAFTING AND REVIEWING A NONPROVISIONAL PATENT APPLICATION

As discussed in previous chapters, the patent process can begin by first filing a provisional application and then subsequently filing a nonprovisional within one year. Alternatively, patent applicants can skip the provisional patent application step and simply file a nonprovisional application first.

Where a provisional patent application was filed first, applicants should begin working on the nonprovisional one to two months before the expiration of the provisional patent application. The first step to preparing the nonprovisional is studying the provisional patent application that was filed and determining any changes, updates or additions that were made to the invention since the provisional application was filed. If there are substantial differences, it may be necessary to draft a new invention disclosure and/or meet with your patent attorney for a new inventor interview to discuss the changes.

Assuming a comprehensive provisional patent application was filed and there are no changes to the invention, then the only updates needed are drafting a set of patent claims and making revisions to specification to better support the newly drafted claims. However, any changes or additions to the invention need to be included in the nonprovisional application by simply adding the new material to the existing application and drafting a set of patent claims.

On the other hand, drafting a nonprovisional patent application first instead of a provisional patent application is essentially the same as drafting a comprehensive provisional patent application, as discussed in Chapters 9 and 10, aside from drafting patent claims and creating an examination strategy. Working with your patent attorney to draft claims for a nonprovisional is effectively the same, regardless of whether a provisional patent application was filed first or not.

Since nonprovisional patent applications are substantively reviewed by a patent examiner, forming a strategy for the examination process begins with how the patent claims are drafted in the nonprovisional application that is filed with the USPTO. Working with an examiner during the examination process should be thought of like the negotiation that occurs during the sale of real estate or other goods. For example, the process of buying a home begins with a buyer finding a house and making an offer. A typical strategy is to make an offer that is lower than the asking price. In such a case, there is an assumption that the lower offer is an initial offer in a negotiation that may go back and forth between buyer and seller several times before an agreement is reached. Often times, the final agreement is for a price that is between the asking price and the initial offer.

In the examination of a patent application, the claims that are initially filed in a nonprovisional are like the initial offer in a negotiation for a house and the scope of the claims is analogous to the price that is offered. For example, just as a home buyer can submit a low or high offer, patent applicants can submit broad or narrow claims. As discussed above, broad claims that have fewer elements are more valuable, and are therefore analogous to making a low offer when buying a house. A low initial house offer will almost certainly result in a negotiation, but is more likely to result in a better final price for the buyer, and if the buyer gets lucky, the initial offer might be accepted. Similarly, by initially submitting broad patent claims, the applicants are more likely to receive some pushback from the examiner, but are more likely to get broader and therefore more valuable claims in an issued patent. On the other hand, initially submitting narrow claims makes it more likely that the examination process will go much faster, but the issued claims will likely be narrower and therefore less valuable than possible.

However, unlike a real estate negotiation, the negotiation for broader claims with a patent Examiner incurs additional cost because a patent attorney must analyze rejections made by the Examiner and draft suitable responses in hopes of getting the application allowed. Accordingly, a negotiation that results in a broader and more valuable patent may be more costly because of the attorney-time used during the examination process.

The mechanics of negotiation over the patent claims that occurs during the examination process is discussed in more detail in the coming chapters along with advanced strategies for the patent process and building a patent portfolio. However, for the time being, it is enough to understand that a nonprovisional patent application can be drafted for faster or slower examination. The patent attorney and applicant should therefore work together to craft a strategy that is appropriate for the unique needs of a given business. For example, where the company would benefit from having a patent issued faster with lower examination costs, it might be beneficial to initially submit narrower claims. On the other hand, where having the broadest claims possible is most important, submitting broader claims for examination may be a best choice. Regardless, this decision should be made before the patent attorney begins to draft the nonprovisional application.

The patent claims should also be tailored to specific infringers or types of infringers. For example, suppose Recline-O-Co manufactures and sells innovative recliner chairs that have cushions and an internal metal mechanism that allows the chair to extend a footrest, recline, and rock back and forth. Patent claims that cover such recliner chairs might have elements that include the internal mechanism as well as the cushions. However, in the furniture industry, there are some manufacturers that only make the internal mechanisms of a chair and sell them to other companies that subsequently add cushions. If these competing companies are making and selling knockoff products, all levels of the supply chain should ideally be covered by Recline-O-Co's patent claims.

The final product would infringe claims that recite elements of both the mechanism and cushions. On the other hand, the company that sells the knockoff mechanism would not infringe claims with elements that recite both the mechanism and cushions. Recall that direct patent infringement occurs only if all elements are present in an accused product. Here, since cushions are not yet present on the mechanism, patent claims that recite cushions would not be directly infringed by the mechanism alone because the cushions element is not satisfied.[1]

The easy solution is to draft some or all of the patent claims so that they only include elements of the innovative mechanism. Such claims would make both the mechanism manufacturer and the manufacturer of the complete chair direct infringers. Again, recall that infringement occurs when all elements of a patent claim are met, and here, the mechanism alone has all the elements and the complete recliner also has all

---

1. The mechanism manufacturers might still be held liable under a theory of indirect or contributory infringement, which is essentially arguing that they should be liable for infringement too because their products are primarily used to make infringing products. However, proving such an infringement theory is much more difficult and often fails. Plus, arguing indirect or contributory infringement makes enforcement substantially more expensive, which is why setting claims up correctly is so important.

the elements. The fact that the final chair has cushions or other parts makes no difference as long as the underlying infringing mechanism is present in the chair.

The specific techniques of drafting patent claims directed to specific infringers are outside the scope of this book because seasoned patent attorneys should be relied on in drafting patent claims to target different types of infringers. On the other hand, patent attorneys are not always intimately familiar with the business structures or supply chains of every industry, and clients should make sure that their patent attorneys have an adequate understanding of the particular industry landscape to draft the patent application appropriately.

REVIEWING A DRAFT OF A NONPROVISIONAL PATENT APPLICATION

The process of drafting a nonprovisional application is essentially the same as drafting a provisional patent application with an attorney, as discussed in Chapter 10. Once the invention has been adequately disclosed to the patent attorney, and once a general strategy for examination and targeting infringers has been established, the patent attorney will proceed with crafting a complete patent application. A draft will be provided for final review and approval by the applicant before preparing the application for filing at the USPTO.

Reviewing the specification and drawings of the patent application is essentially the same as reviewing the specification and drawings of a provisional application as described in Chapter 10—the focus should be on confirming that the invention is disclosed correctly, with sufficient detail, and that alternative embodiments are also adequately described.

With a more developed understanding of patent claims, the reason for this focus becomes clearer. For example, consider the drawings and specification to be like a food buffet. For those of you who have ever had the pleasure of eating at a large buffet, you know the delight of being able to make a plate of food that is specifically tailored to your appetite at that moment. You can have lasagna with chow mein and a hamburger, which all sits next to marshmallows that have been dipped in a chocolate fountain. The possibilities are endless.

Similarly, the drawings and specification of a patent application are the buffet that allows for crafting the perfect plate of patent claims. As discussed in coming chapters, the goal of the examination process is to find a set of patent claims with just the right elements so that the examiner finds them allowable, but with as few elements as possible, so that the claims remain broad. However, this plate of patent claims is limited by the buffet of elements that are included in the specification and drawings at the time of filing. In other words, the plate of claims cannot include elements that are not described in the drawings or specification, and missing elements cannot be added after the application is filed.

Having a wide variety of elements to choose from in the specification and drawings is important because it is impossible to know what elements will later be critical to the claims at the time the patent application is filed. For example, given how unpredictable examiners and the examination process can be, it is impossible to know what specific set of claim elements might be needed to reach the threshold of patentability. Having a variety of elements in the specification and drawings to choose from is therefore essential.

Additionally, as described in more detail in coming chapters, an advanced patent strategy is to draft patent claims to directly read on a competitor product that comes out after the patent application is initially filed. However, to do this, elements that describe this new product must already be described in the patent specification or drawings. Accordingly, a patent application should not only describe the invention at hand, but also attempt to describe alternative embodiments or related inventions that might be developed by competitors in the future. Patent attorneys try to add as many alternative embodiments to the application as possible, but applicants should review the application to make sure that no important alternatives have been left out or incompletely described.

As for the claims, applicants should read and ask their patent attorney to briefly explain the claim strategy being implemented. Receiving a master class in claim drafting is not an efficient use of a patent budget, but applicants should confirm that an examination and infringement target strategy has been implemented correctly as discussed above.

Additionally, I strongly recommend against editing the claims in a patent application draft. Each word that is included and *not* included in a patent claim is intentional, and even minor changes can harm the overall structure and coherency of a set of claims. Moreover, USPTO rules, patent laws, and years of case precedent often dictates specific sentence structures and word choices that can make a critical difference in the examination process and in the subsequent enforcement of an issued patent.

Instead, it is more efficient to formulate questions about the claims before making any edits or additions. Applicants should treat the claims almost like a foreign language, and think of their patent attorney as a translator for this strange language that often resembles English. Doing so will help keep the patent application cost within budget while also assuring the highest quality possible.

FINAL STEPS BEFORE FILING

With a final draft of the patent application reviewed and approved, a few steps remain before the patent application is ready to be filed at the USPTO. The majority of these steps are the same as when filing a provisional patent application as discussed in detail in Chapter 10, including identifying the inventors and executing an assignment of the patent application where applicable.

If a provisional patent application was previously filed, keep in mind that the inventors named in the nonprovisional application might be different. For example, where changes or additions to the invention were made since the provisional application, inventors of the new aspects of the invention might need to be named. Additionally, inventorship in a nonprovisional application is defined by what is described in the claims and not by what is described in the specification and drawings. The claims include only a small subset of what is described in the overall application, and if there are inventors of subject matter that is described in the specification or drawings but not included in the claims as filed, then these persons should not be named as inventors in the nonprovisional application. If it is unclear who should be named as an inventor, collaborate with your patent attorney to work it out before the application is filed. Naming the wrong set of inventors can lead to invalidation of a patent and loss of control over patent rights. At the very least, naming the wrong inventors up-front incurs additional unnecessary cost down the road.

Along with the filing steps that are the same as a provisional application, there are several new documents that may need to be filed along with the application, or shortly thereafter if time is limited. The inventors and other persons associated with the patent application, including company officers and even the patent attorney, are required to disclose known relevant prior art to the USPTO. This requirement is fulfilled by filing an information disclosure statement (IDS), that lists known prior art and requires submission of prior art documents that are not issued U.S. patents or U.S. patent publications.

Intentional failure to submit these prior art references to the USPTO can result in invalidation of an issued patent, so this duty to disclose should be taken very seriously. Applicants can help reduce costs related to drafting an IDS and compiling documents for submission by providing a list of all known prior art references, and a copy of all documents that are not related to United States patents or applications. Ideally, all of these documents should be given to your patent attorney in electronic form as a PDF document, which is the format in which they will be submitted to the USPTO. Providing this list and documents prevents cost associated with attorneys or billable staff having to hunt for and procure them.

In nonprovisional applications, all inventors must sign an official declaration form that confirms their status as an inventor in the application and acknowledges that willful false statements of inventorship are punishable by a fine or imprisonment. A simple signature of each inventor is all that is required. Recent rule changes allow inventor declarations to be submitted after the application is filed and up until the application is allowed, but declarations should be executed by inventors as soon as possible. It can be several years until the final deadline for submitting inventor declarations, and by this time the inventors may not be associated with the company anymore and may be difficult to reach.

Another form that should be submitted at filing of the nonprovisional patent application is a power of attorney form. This form gives your patent attorney official authority to handle the case and directs that mail from the USPTO go to the attorney instead of the applicant or one of the named inventors. This is standard practice and should always be done. It prevents loss of correspondence as company or personal addresses change and allows the attorney to efficiently attend to issues that may arise during the examination of the application.

By default, a nonprovisional patent application will be published and the application contents will become publicly available 18 months from the first priority date. For example, if a provisional application was first filed, and a nonprovisional application was filed exactly one year later, the nonprovisional application will become public approximately six months later. Alternatively, if the nonprovisional application is the first patent filing, it will become public after about 18 months.

However, this default can be changed. Applicants can opt for early publication, which makes the nonprovisional application publicly available soon after the request is made. Alternatively, if patent protection is not being sought in foreign countries, applicants have the option of nonpublication, which keeps the nonprovisional application secret until it issues as a patent. The 18-month default publication option is the most common, but if early publication or complete secrecy is desired, such a request should be made at the time of filing.

These associated documents and the patent application are then filed with the USPTO electronically and a filing receipt is immediately generated that confirms safe receipt of the patent application and provides an official filing number. The invention continues to enjoy "patent pending" status if a provisional was previously filed, or gains "patent pending" status immediately when the nonprovisional is the first patent application filed.

POST-FILING CONSIDERATIONS

Once a nonprovisional application is filed, it may be one to three years before the examination process begins. However, there is still plenty of work to be done before the application is eventually picked up by an examiner. For example, just as with a provisional application, the invention is "patent pending" once the nonprovisional application is filed, and applicants should feel comfortable making, using, and selling the technology. There is no reason to wait until the patent issues to start exploiting the invention. Keep in mind, however, that only the technology described in the application is "patent pending," and if substantial changes or additions are made to the invention, a new application may need to be filed.

In fact, it is a good idea to periodically check in with your patent attorney while waiting for the examination phase to make sure that no changes need to be made to your patent strategy. Changes to business strategy or structure, along with changes in the marketplace, may necessitate a modified plan. For example, if competitors start to copy or make products that are similar to the invention in the patent application, it might be worthwhile to amend the pending patent claims to directly read on this new product, if possible. Alternatively, where the commercial embodiment of the invention changes compared to the product that was sold at the time of filing, it could be beneficial to amend the patent claims to better cover the new commercial product. Alternatively, it may even be necessary to file another application that includes disclosure of this new product. Although patent applications attempt to broadly cover an invention so that many variations are included under the patent, some modifications or additions will not be adequately protected by the original application. Unfortunately, it is not uncommon for issued patents to not even cover the commercial embodiment of an invention, which is often a result of lack of communication between the patent attorney and applicant.

Additionally, the duty to disclose known prior art to the USPTO continues from the time of filing until the application issues as a patent. Accordingly, if relevant prior art becomes known to the inventors or others involved with the patent application, this prior art should be shared with your patent attorney so that it can be submitted to the USPTO in an IDS. Again, failure to do so can result in later invalidation of a patent and can cause problems for enforcement.

Unless a nonpublication request is filed, the application will publish 18 months from the earliest priority date. This can be an important marketing opportunity for a company, and a patent publication announcement can be aligned with other marketing efforts. On the other hand, if the needs of a company change, a nonpublication request or early publication request can be filed before the default publication occurs.

After one to three years, a patent examiner will analyze the filed patent claims to determine if they are patentable. Rejection should be expected as a normal part of the examination process at the USPTO, and the following chapter discusses how to interpret and respond to such rejections with the assistance of a patent attorney.

CHAPTER 11 SUMMARY

- The claims of an issued patent define a specific set of elements that the patent owner can exclude others from making, using, offering for sale, or selling throughout the United States or importing into the United States.
- Depending on the elements recited in the claims, patents can be difficult to infringe or easier to infringe (i.e., patents can be broad or narrow).
- Patent infringement occurs when a court finds that a product or process includes all recited elements of at least one claim in an issued patent.
- In contrast, patentability is judged by a patent examiner at the USPTO who determines whether the pending claims in a patent application are new and nonobvious in view of the prior art.
- The examination process is like a negotiation with a patent examiner where the goal is to present a set of patent claims that are as broad as possible (i.e., have maximum value) while also being new and nonobvious in view of the prior art.
- The specification and drawings of a top-quality provisional application should be very similar to those of a nonprovisional application, but the patent claims that are required in nonprovisional applications are typically omitted in provisional applications.
- When reviewing a draft of a nonprovisional application, applicants should make sure they understand the claim strategy, but given their technical nature, applicants should avoid trying to edit the patent claims.

12

The Examination Process: Deciphering Office Actions and Helping Your Attorney Respond

"The Director shall cause an examination to be made of the application and the alleged new invention; and if on such examination it appears that the applicant is entitled to a patent under the law, the Director shall issue a patent therefor."

—Title 35 of the United States Code, Section 131

After a nonprovisional patent application is filed at the USPTO, it is initially processed to make sure that the basic application requirements have been met. The application is categorized by type of invention and is assigned to an "art unit" that will handle the application examination. Each USPTO art unit specializes in examining certain types of inventions, and is staffed with patent examiners who have technical backgrounds relevant to the types of inventions they examine. Patent applications are examined on a first-come, first-served basis, which accounts for the long wait from filing to the beginning of the examination. Some art units have a larger backlog than others. For example, computer hardware and software applications typically have the longest wait time—often two to three years. On the other hand, mechanical devices often have a shorter wait—sometimes one to two years.

Each application is assigned to a patent examiner, who will be solely responsible for the examination process. When an application comes to the top of the pile, the examiner first determines whether the claims are directed to more than one invention. If so, the applicant will be required to select one invention for examination. The examiner then reviews the patent claims and does a prior art search to determine if the invention defined by the patent claims is new and nonobvious compared to the prior art. Other less-substantive formality requirements may also be addressed.

The examiner will draft a formal document, called an Office Action, which sets forth rejections that the examiner believes prevent the application from being allowable. Applicants then have the opportunity to amend the patent claims and/or argue against the rejections. In certain limited circumstances, the drawings or specification may receive minor edits as well. The patent attorney representing the case will draft a formal response that sets out any arguments and amendments made to the application, and this response is typically filed within two or three months of when the Office Action was received.

This back-and-forth will continue until the examiner allows the application, or the applicant gives up and abandons the application. Alternatively, if the applicant disagrees with the rejections and the Examiner is unwilling to compromise, the applicant can appeal the case to an appeals board at the USPTO and ultimately to federal courts if necessary. If the examiner allows the application, final issue fees are paid to the USPTO and the patent application will issue as an enforceable patent shortly thereafter.

When preparing and budgeting for the examination process, applicants should expect application rejections as a natural part of getting a patent issued. Again, the examination process is like a negotiation, and

Examiner rejections along with attorney responses should be considered analogous to offers and counter-offers with the goal of getting the broadest claim scope possible. The nuanced and formal dance of patent examination is a skill that is slowly honed through years of experience, and applicants should largely rely on their patent attorney to craft a strategy for getting a patent allowed as quickly as possible while also maximizing claim-scope. However, the right applicant input is essential to assuring that the patent attorney has the necessary information to move the application forward, while ensuring that the patent strategy still meets the needs of the company. Accordingly, the following sections provide insight into deciphering the hidden meaning of Office Actions received from an examiner, and offer insider tips on providing valuable analysis and input to the examination process without unduly increasing the cost.

ACCESS TO PATENT EXAMINATION DOCUMENTS

All documents in published patent applications are publicly available via the USPTO's online Patent Application Information Retrieval (PAIR) system. By providing an application number, a publication number or a patent number, applicants and others can monitor the status of pending cases and review the complete history of filed documents. Unpublished applications remain secret, however, and are not open for public review.

SUBSTANTIVE REJECTIONS—NOVELTY AND OBVIOUSNESS

There are many types of rejections that an examiner might raise against a patent application, but the most substantive are based on the novelty or obviousness of the claimed invention. Other types of rejections are less important because they can almost always be overcome by making minor changes to the claim language. When patent applications fail to issue, however, it is nearly always because the applicant was unable to overcome novelty or obviousness rejections asserted by the examiner.

NOVELTY REJECTIONS—35 U.S.C. §102

After performing a prior art search, the examiner will compare the prior art to the patent claims and determine whether the claim language is novel. To make this determination, the examiner will compare the elements of a claim to each prior-art reference and see if all of the claim elements are taught by the reference. For example, if a prior art reference is an issued patent, the examiner will search through the specification and drawings of the patent and see if all claim elements are described or shown.

> Claim 1: A vehicle comprising:
> (A) a front and rear wheel;
> (B) a frame including:
> (D) handlebars; and
> (F) at least two bars spanning between the wheels; and
> (I) a pedal assembly coupled with the rear wheel via a chain.

To better understand this, recall the example patent claims in Chapter 11 that had a set of elements that described a bicycle invention. One such bicycle claim, as illustrated above, recites a "vehicle comprising: a front and rear wheel; a frame including handlebars and at least two bars spanning between the wheels; and a pedal assembly coupled with the rear wheel via a chain." Let's further arbitrarily break this claim down into a set of five major elements, which as labeled above are elements A, B, D, F, and I.

Now consider that a patent examiner did a prior art, and discovered two relevant prior art references that were similar to the invention defined by claim 1 above. The first prior art reference teaches a bicycle having a plurality of elements designated A, B, C, D, E, F, G, H and I. The second reference teaches a plurality of elements designated A, B, C, D, E, F, G and H. For the first prior art reference, the examiner would compare all of the elements of claim 1 to what is described in Prior Art 1. The examiner would find that Prior Art 1 teaches elements A, B, D, F and I that are the same elements as in claim 1, in addition to teaching elements C, E, G and H, which are not in claim 1. However, because all the elements of claim 1 are taught, the examiner could reject claim 1 for lacking novelty in view of Prior Art 1.

Claim 1:	A		C	D		F				I
Prior Art 1:	A	B	C	D	E	F	G	H	I	
Prior Art 2:	A	B	C	D	E	F	G	H		

On the other hand, when comparing Prior Art 2 to claim 1, the examiner would find that elements A, B, D, and F are taught by Prior Art 2, but that element I is not taught by Prior Art 2. The examiner would therefore not be able to reject claim 1 for lacking novelty in view of Prior Art 2, because element I is not taught or suggested by Prior Art 2.

In view of this analysis, the examiner would reject a patent application that had claim 1 and draft an Office Action that set forth this rejection. For example, such a rejection is often worded as follows:

CLAIM REJECTIONS—35 U.S.C. § 102

1. The following is a quotation of the appropriate paragraphs of 35 U.S.C. 102 that form the basis for the rejections under this section made in this Office Action:

continued on next page

A person shall be entitled to a patent unless—

(b) the invention was patented or described in a printed publication in this or a foreign country or in public use or on sale in this country, more than one year prior to the date of application for patent in the United States.

2. Claim 1 is rejected under 35 U.S.C. 102(b) as being anticipated by [Prior Art 1]. With regard to independent claim 1, [Prior Art 1] discloses a vehicle comprising: a front and rear wheel [cite 1]; a frame [cite 2] including handlebars [cite 3] and at least two bars spanning between the wheels [cite 4]; and a pedal assembly coupled with the rear wheel via a chain [cite 5].

In a real Office Action, the examiner will cite the prior art reference being used for the novelty rejection, which is typically an issued patent or publication of a patent application, and will also cite to specific portions of the prior art reference where the examiner argues the claim elements are disclosed. Additionally, patent applications usually have more than one claim, so the examiner might also set forth rejections of the other claims along with other types of rejections. To view examples of what real USPTO Office Actions look like for various types of inventions visit www.PatentsDemystified.com.

For a patent application to be allowable, the examiner must agree that all claims are new and nonobvious in view of the prior art, and in this example Office Action, because claim 1 is rejected, the application would not be in condition for allowance. In response to this rejection, applicants would have two primary options—they can amend the claim so that it includes at least one element that is not taught by the prior art reference used in the rejection and/or they can argue against the rejection in hopes of changing the examiner's mind.

With a limited amount of time to examine each application, it is not uncommon for examiners to misinterpret claim elements, misunderstand what a prior art reference actually discloses, or even leave out claim elements from their analysis. In these cases, applicants may be able to respond to the rejection without making amendments to the claims.

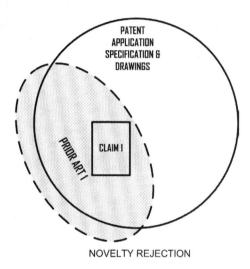

NOVELTY REJECTION

FIGURE 12.1

Where an examiner makes a solid argument, however, it may be necessary to amend some patent claims to include at least one element that is not taught in the prior art reference that was used in the novelty rejection. For example, consider the Venn diagram in Figure 12.1 that illustrates a novelty rejection. The scope of the patent application specification and drawings is represented by the large circle, and claims are required to be within the scope of the patent specification and drawings. Here, the scope of claim 1 is illustrated by the rectangle, and the scope of the prior art reference used to reject claim 1 is shown by the shaded oval. Notice that the prior art reference discloses some things that are outside the scope of the patent application documents, and also discloses some things that are outside the scope of claim 1. These portions outside the rectangle are irrelevant to the rejection of claim 1. Instead, the core of the examiner's rejection of claim 1 is that some portion of the prior art reference covers all elements in claim 1.

To make claim 1 allowable, elements must be added that are not taught by the prior art reference, but that are already described in the patent application specification. In the Venn diagram in Figure 12.2, the rectangle that represents claim 1 is now larger and includes a small corner that is not within the scope of the prior art reference. This amendment includes at least one new element that is not taught by the prior art reference, which effectively overcomes the novelty rejection presented by the examiner in the Office Action.

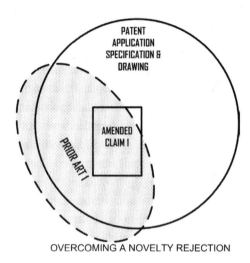

OVERCOMING A NOVELTY REJECTION

FIGURE 12.2

Referring to the example above where claim 1 was considered to have elements A, B, D, F and I, assume that claim 1 is amended to include new element J shown in bold. The prior art reference still has elements A, B, D, F and I, but it does not teach newly added element J. Accordingly, by adding element J, claim 1 overcomes the novelty prior art rejection that was based on Prior Art 1. Importantly, this illustrates that claims only need to have one part that is new compared to the prior art to overcome a novelty rejection, and there is no requirement that all elements recited in a claim be novel. This element that tips the scales into patentability is often called the point of novelty.

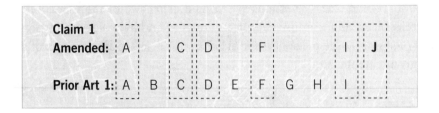

Additionally, recall that less is more for patent claims and that having fewer elements is better because it makes it easier to infringe and therefore make the claim broader and more valuable. Accordingly, applicants should amend claims only the minimal amount required to make them novel in view of asserted prior art. In some cases, simply adding a single word can be enough to overcome a rejection, whereas in other cases, several lengthy elements might be required to make a claim novel. The subtle art of maximizing claim scope during examination is what makes a skilled patent attorney so valuable, and applicants should allow their patent attorney to plan and execute an the examination strategy. However, as discussed in more detail later in this chapter, applicants can nonetheless play a crucial role in analyzing prior art and patent claims, and can provide their patent attorney with the information needed to overcome tough rejections.

Unfortunately, overcoming all of the rejections set forth in an Office Action does not guarantee that the patent application will be allowable. In response to claim amendments and/or arguments against rejections, the examiner might concede that the applicant is right, but might conduct another prior art search and try to find prior art that teaches a new element that was added to the claim. In a subsequent Office Action, the examiner might present a new novelty rejection or might come up with an obviousness rejection of the claim.

OBVIOUSNESS REJECTIONS—35 U.S.C. §103

Another requirement for getting a patent allowed is that the claimed invention is not obvious in view of the prior art. More specifically, if a hypothetical person with ordinary skill in the art were aware of all available prior art, the patent claims cannot have a set of elements that would be obvious to this hypothetical person. Unlike novelty rejections, obviousness rejections are formed using two or more prior art references, and examiners typically turn to an obviousness rejection when one single piece of prior art does not disclose all elements of a given patent claim.

Claim 1:	A	B		D		F			I
Prior Art 3:	A	B	C						
Prior Art 4:	A		C			F	G	H	I
Prior Art 3+4:	A	B	C	D		F	G	H	I

For example, using the same example patent claim as above that has elements A, B, D, F, and I, consider that the Examiner finds two alternative pieces of prior art. A novelty rejection cannot be formed from Prior Art 3 because it only has elements A and B—elements D, F and I are missing. The examiner is also unable to form a novelty rejection from Prior Art 4 because it only has elements A, F, and I—elements B and D are missing. However, if the elements of Prior Art 3 and 4 are combined, then all elements A, C, D, F and I of claim 1 are taught. The examiner is therefore arguing that it would be obvious to combine these prior art references, and that doing so collectively satisfies all elements of claim 1.

To better understand this obviousness rejection, consider the Venn diagram in Figure 12.3, where the large circle represents the patent application specification and drawings and the rectangle represents claim 1. The ovals that illustrate Prior Art 3 and 4 both include some elements that are not in the specification or claims; some elements that are in the specification, but not the claim; and some elements that are in both the specification and the claim. However, neither oval fully covers claim 1, which indicates that neither Prior Art 3 nor 4 has all elements of claim 1. Additionally, there is some overlap between Prior Art 3 and 4 in claim 1, which indicates that both references include some of the same elements. Regardless, the crux of this obviousness rejection is that Prior Art 3 and 4 collectively cover all elements of claim 1.

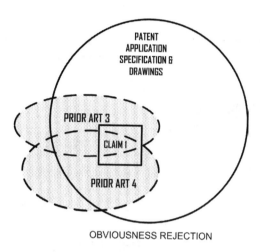

OBVIOUSNESS REJECTION

FIGURE 12.3

In view of this analysis, the examiner would reject the patent application that had claim 1 and draft an Office Action that set forth this obviousness rejection. For example, such a rejection is often worded as follows.

CLAIM REJECTIONS—35 U.S.C. § 103

1. The following is a quotation of 35 U.S.C. 103, which forms the basis for all obviousness rejections set forth in this Office Action:

A patent for a claimed invention may not be obtained, notwithstanding that the claimed invention is not identically disclosed as set forth in section 102, if the differences between

continued on next page

the claimed invention and the prior art are such that the claimed invention as a whole would have been obvious before the effective filing date of the claimed invention to a person having ordinary skill in the art to which the claimed invention pertains. Patentability shall not be negated by the manner in which the invention was made.

2. Claim 1 is rejected under 35 U.S.C. 103 as being unpatentable over [Prior Art 3] in view of [Prior Art 4].

3. With regard to independent claim 1, [Prior Art 3] discloses a vehicle comprising: a front and rear wheel [cite 1]; a frame [cite 2].

4. [Prior Art 3] does not teach a frame including handlebars and at least two bars spanning between the wheels; and a pedal assembly coupled with the rear wheel via a chain. However, in the same field of endeavor, [Prior Art 4] discloses a frame including handlebars [cite 3] and at least two bars spanning between the wheels [cite 4]; and a pedal assembly coupled with the rear wheel via a chain [cite 5].

5. It would have been obvious for one of ordinary skill in the art at the time of the invention to modify the wheels and frame of [Prior Art 3] to include handlebars and at least two bars spanning between the wheels; and a pedal assembly coupled with the rear wheel via a chain as taught by [Prior Art 4].

Just as with novelty rejections, applicants can overcome an obviousness rejection by arguing to change the examiner's mind and/or by amending the rejected claims to include at least one element that is not obvious. Arguments against obviousness rejections can include arguments that the examiner misinterpreted the claim language or the teachings of the prior art, as with novelty rejections, but there are other ways to attack an obviousness rejection. For example, applicants can argue that the two or more references that make up the rejection cannot be properly combined, or even argue that the claims are not obvious because the invention has had commercial success, satisfies a long-felt and unsolved need or that others have failed when attempting to come up with the claimed invention. However, the easiest and most common way to overcome an obviousness rejection is essentially the same as overcoming a novelty rejection— applicants can amend the claim to include an element that is not taught or suggested by the two or more prior art references that make up an obviousness rejection.

Claim 1 Amended:	A	B		D		F				I	J
Prior Art 3:	A	B	C								
Prior Art 4:	A		C			F	G	H	I		
Prior Art 3+4:	A	B	C	D		F	G	H	I		

For example consider amended claim 1 above and the Venn diagram in Figure 12.4, which both depict adding new element J to claim 1. As before, elements A, B, D, F and I are covered by Prior Art 3 and 4 if these references are combined, but neither piece of prior art discloses element J, which will be sufficient to overcome the obviousness rejection of claim 1 in view of Prior Art 3 and 4. As shown in the example Venn diagram, the rectangle has been expanded to the right, which represents the addition of element J to claim 1. The rectangle of claim 1 now includes a small portion that is not covered by Prior Art 3 or 4, which is sufficient to overcome the obviousness rejection.

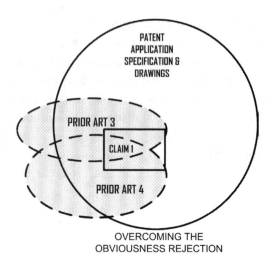

OVERCOMING THE
OBVIOUSNESS REJECTION

FIGURE 12.4

However, with both obviousness and novelty rejections, keep in mind that the claims can only be expanded within the scope of what is already described in the specification and drawings of the patent application. They cannot be amended to include elements that are not in the specification and drawings and the specification and drawings cannot be amended to add new elements after the application is filed. This illustrates why it is important to have a broad disclosure in the patent specification and drawings at the time of filing. The Venn diagram in Figure 12.5 illustrates a patent application where claims 1 and 2 have been rejected by an examiner in view of prior art references 5 and 6. Claim 1 has a novelty rejection because its elements are solely covered by Prior Art 5. On the other hand, some elements of claim 2 are covered by one of Prior Art 5 and 6 alone, while other elements are covered by both Prior Art 5 and 6. Claim 2 is therefore rejected based on obviousness in view of the combination of Prior Art 5 and 6. Unfortunately, there would likely be no way to overcome the rejections in this example because regardless of how claim 1 or 2 is expanded, there is no part of the specification and drawings that is not covered by one or both of Prior Art 5 and 6. The circle that represents the specification and drawings cannot be expanded because adding new matter to an application after filing is grounds for rejection by the examiner. Alternatively, expanding either claim 1 or claim 2 outside of the circle of the specification and drawings would likewise be cause for rejection.

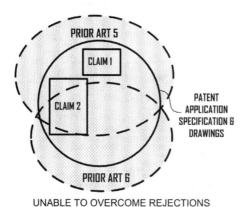

UNABLE TO OVERCOME REJECTIONS

FIGURE 12.5

Although the mechanics, strategy and specific language of claim amendments and arguments should be left to the patent attorney, applicants should still understand the theory behind these rejections and the ramifications of making claim amendments and/or arguing against rejections. However, applicants can help their attorney while also saving money and improving the quality of the claims by properly assisting with analysis of both the asserted prior art references and the patent specification and drawings.

ASSISTING WITH AN OFFICE ACTION RESPONSE

All correspondence from the USPTO, including Office Actions, is sent directly to the patent attorney, who will then send it to the applicants for review. A formal response must typically be filed within two or three months of the Office Action mailing date, and up to six months from the Office Action mailing date if extension-of-time fees are paid. The patent attorney can draft a response without applicant input, but applicants can often make valuable contributions that will likely improve the quality of the response and possibly reduce the preparation costs.

Novelty and obviousness rejections provide the best opportunity for applicants to assist with the relevant analysis. Other types of rejections are either non-substantive or require the knowledge of specific patent laws or rules and it is more efficient for the patent attorney to handle these rejections independently. In the claim examples above, only one or two claims were considered at a time. However, a typical patent application includes around 20 claims of different types that are interrelated in various ways as discussed below.

THE TYPES OF CLAIMS

Patent claims come in two structural forms—independent and dependent claims. Claim 1 in the example set of three claims below is an independent claim much like the independent claims discussed above in relation to novelty and obviousness rejections. In this example, the claim has five elements labeled A to E. (Note that the parenthetical labels are not present in real claims and are only presented here to mark the different elements).

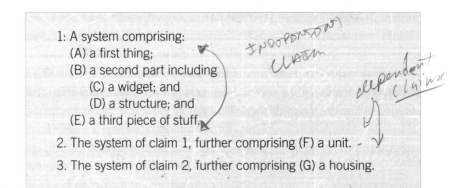

1: A system comprising:
   (A) a first thing;
   (B) a second part including
     (C) a widget; and
     (D) a structure; and
   (E) a third piece of stuff.
2. The system of claim 1, further comprising (F) a unit.
3. The system of claim 2, further comprising (G) a housing.

Claims 2 and 3 are dependent claims because they "depend" from or cite reference to another claim. In this example, claim 2 depends from claim 1 and claim 3 depends from claim 2. As discussed above, the scope of a claim is defined by its elements. For independent claims like claim 1, this is simply elements A through E in the body of the claim. For dependent claims, however, the scope is defined by both the elements in the dependent claim *and* the elements of the claim that it depends from. For example, because claim 2 depends from claim 1, the scope of claim 2 is elements A through E and F. Claim 3 depends from claim 2, which depends from claim 1, so claim 3 is defined by elements of all three claims—namely elements A through E, F, and G.

Recall that less is more when it comes to patent claims, so independent claim 1 is the broadest and best claim because it has the least number of elements (only A through E). Claim 2 is narrower because it has one more element than claim 1, and claim 3 is narrower than claims 1 and 2 because it has the most elements. In fact, by definition, independent claims *must* be the broadest claims and dependent claims *must* be narrower than independent claims. This nested relationship is illustrated in the Venn diagram in Figure 12.6.

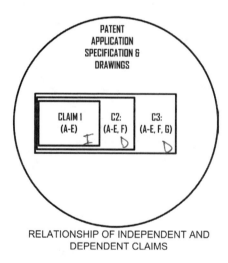

RELATIONSHIP OF INDEPENDENT AND
DEPENDENT CLAIMS

FIGURE 12.6

Having claims of different scopes is important to the strategy of the examination process and provides the patent examiner with a range of options where at least one set of elements will hopefully be found to be new and nonobvious over the prior art. For example, as discussed above, having at least one element in a claim that is not taught by any prior art reference provides a way to argue against novelty or obviousness rejections of that claim or can prevent the claim from being rejected in the first place.

Using the example set of three claims above, assume that the examiner does a prior art search and finds a piece of prior art that has the scope illustrated in the Venn diagram in Figure 12.7. Prior Art 1 covers all of claim 1 that has elements A-E and also covers all of claim 2 that has elements A-E and F. The examiner would therefore reject claims 1 and 2 for not being novel in view of Prior Art 1.

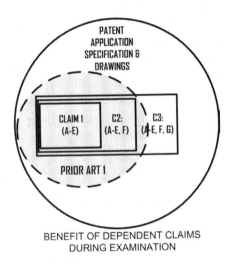

BENEFIT OF DEPENDENT CLAIMS
DURING EXAMINATION

FIGURE 12.7

However, part of claim 3 is not covered by Prior Art 1, and thus the examiner would not be able to make a novelty rejection of claim 3 based on Prior Art 1. Assuming there are no other rejections that apply to claim 3, the examiner would then indicate in an Office Action that the elements of claim 3 would be allowable. This all but guarantees that the patent application will be allowed to issue with only a few minor edits to the claims. Moreover, it provides insight that element G is the point of novelty that tips the scales into the realm of patentability.

With knowledge that elements A-E, F, and G combined would be allowable, one option is to include the language of claims 2 and 3 into claim 1, so that all these allowable elements are present in one claim. Claims 2 and 3 would then be canceled. Additionally, if the application includes any other claims that depend from claim 1, these dependent claims would also be automatically allowable because claim 1 is now allowable. Unless there are other rejections against this amended claim, or other claims, the examiner will allow the application based on these relatively simple amendments.

> 1: (Amended) A system comprising:
> (A) a first thing;
> (B) a second part including
>   (C) a widget; and
>   (D) a structure;
> (E) a third piece of stuff;
> (F) a unit; and
> (G) a housing.
>
> 2. (Canceled)
>
> 3. (Canceled)

These examples illustrate how important it is to find even one element that is not taught or suggested by prior art asserted by the examiner, because finding such an element can quickly lead to allowance of the patent application. Accordingly, proper analysis of an Office Action, cited prior art, and the specification and drawings of the patent application are required to secure allowance of the application.

ANALYSIS TO FIND PATENTABLE ELEMENTS

Before embarking on a time-consuming analysis of claim rejections, prior art, and the patent application drawings and specification, confer with your patent attorney to confirm that such an analysis is necessary. In many cases, some claim elements might be indicated as allowable as discussed above, or the examiner may indicate other relatively simple amendments that will make the patent application allowable. However, where simple amendments or arguments to overcome novelty or obviousness rejections are not immediately clear, it becomes necessary to do a more in-depth analysis, to which applicants can make a valuable contribution. The trick is providing your attorney with analysis that is relevant to responding to rejections and moves the response drafting process forward, but does not incur unnecessary attorney time. The right contribution, in a streamlined format, can substantially reduce the cost of responding to an Office Action.

As discussed above, a primary goal of the patent examination process is to find at least one element of the invention that is not taught or suggested by any of the asserted prior art references. Once such an element is identified, and assuming the examiner agrees or does not find new prior art that teaches the element, it becomes a simple task to get a patent application allowed. A point of novelty that moves an application into the realm of patentability can be found in two places—in the existing claims or in the patent specification or drawings.

The first step in any analysis is to confirm that the examiner's rejections are reasonable, or at least not worth arguing against. With a large number of cases on their docket and an extremely limited time allocated to each case, patent examiners are typically unable to read and fully comprehend the entire patent application as well as each prior art reference discovered in a prior art search. In fact, time-constrained examiners often read only the patent claims of an application and then briefly review the specification and drawings to better understand the claims before doing a patent search. Rejections for each claim element might be based only on similar keywords in a prior art reference and not on what is actually described in the context of the whole document. Accordingly, for over-burdened patent examiners, the strategy is typically to formulate superficial rejections as quickly as possible and make applicants do the hard and time-consuming work of comparing each of the prior art references to each individual claim element.

Analysis of an Office Action begins by identifying which claims are associated with each obviousness and/or novelty rejection and then making a claim chart for each. For example, consider that claims 1 through 3 discussed above are rejected in an Office Action based on prior art references 1 and 2. A claim chart for this rejection compares the elements of claims 1 through 3 to prior art references 1 and 2.

Setting up the claim chart begins by listing all of the claims in the first column of a table and splitting them into their constituent elements as shown in the table. Independent claims have a plurality of elements that must be split into individual rows. Dependent claims often have only one element, but not always. Real claims will not have the helpful elements label that are shown in these example claims, but line spaces, indents, and punctuation of properly written claims can act as a guide. Regardless, there is no right or wrong way to break a claim down into specific elements as long as the splitting allows for suitable analysis.

	Prior Art 1	Prior Art 2
1: A system comprising:		
(A) a first thing;		
(B) a second part including		
(C) a widget; and		
(D) a structure; and		
(E) a third piece of stuff.		
2. The system of claim 1, further comprising (F) a unit.		
3. The system of claim 2, further comprising (G) a housing.		

Claim chart for analysis of obviousness rejection of claims 1 through 3 based on prior art references 1 and 2.

For obviousness rejections, there will almost always be two or more prior art references, and a column header for each prior art reference goes to the right of the column that has claim elements. The next step is to read through each prior art reference and determine if all elements of the rejected claims are taught or suggested by any part of the prior art reference. Using the broadest reasonable interpretation of each claim element, indicate whether the element is present, is not present, or is possibly present in the corresponding prior art reference. Focus on elements where none of the prior art seems to teach or suggest that element because this can create an argument for that claim being allowable. Novelty rejections are analyzed essentially the same way, except only one prior art reference is compared to the claim elements instead of two or more.

In cases where it seems like all claim elements are taught or suggested in the asserted prior art, the next step is to identify aspects of the invention that are adequately disclosed in the patent application, but are not taught or suggested by the prior art references at hand. These are elements that could be added to the claims, which could possibly move the application into condition for allowance.

Keep in mind that the purpose of preparing these documents is to identify possible arguments or claim amendments that might move the examination process forward. However, just because an argument or claim amendment is possible, does not mean that using it is a good strategy or that it will likely be persuasive to an examiner. In some cases, a given argument or amendment is better saved for later stages of the examination process or as a last resort when the process becomes more difficult. Experienced patent attorneys are experts at crafting Office Action responses and choosing the most appropriate arguments and amendments, so do not be disappointed if every argument or amendment identified in an analysis does not find its way into a given Office Action response.

OTHER REJECTIONS AND OBJECTIONS FOUND IN OFFICE ACTIONS

In addition to obviousness and novelty rejections, there are a variety of other rejections that might arise in an Office Action. Although many of these rejections sound ominous, they are usually easily dealt with by making simple edits to patent claims and sometimes to the drawings or specification. Moreover, such rejections can be purely technical or legal in nature and do not benefit from substantive input from applicants.

DESIGN PATENT EXAMINATION

Because the drawings of design patent applications are analogous to the claims of utility patent applications, the negotiation that occurs during the examination of design applications is centered on the drawings. Although most design applications issue quickly without much substantive examination, some cases can require a few rounds of back-and-forth before the drawings are allowed by the examiner.

For example, the examiner may object to the specification, drawings or claims because of formalities such as typographical errors and other purely technical reasons that involve non-compliance with various USPTO formatting rules. Other times, such formalities can be presented as claim rejections for lack of clarity (cited as rejections under 35 U.S.C. §112). Given that each examiner selectively and sometimes arbitrarily enforces formal patent formatting rules and regulations, such formality rejections appear frequently in patent applications and are nothing to be alarmed about.

Another seemingly serious rejection is for claims that are not directed to patentable subject matter and are not in compliance with 35 U.S.C. §101. The line between patentable and unpatentable subject matter is typically only a matter of adding specific language to the claims that has been determined to be allowable by various court cases. These rejections come up most often in computer-related patent applications in the form of a rejection for claims being directed to an unpatentable abstract idea.

The rules of patent examination dictate that only one invention can be examined per patent application. In other words, the patent claims cannot be directed to more than one invention and cannot create an undue burden on the patent examiner doing a patent search. The line that defines multiple inventions and undue burden is highly subjective and applied arbitrarily from one patent examiner to the next. However, a general rule of thumb is that applications that have more than three independent claims and more than 20 claims total are more likely to be subjected to a restriction requirement that eliminates multiple inventions. Behind the scenes, to make their job easier, over-burdened patent examiners often use restriction requirements to reduce the number of total claims in a patent application.

Restriction requirements typically occur at the beginning of the examination process before the examiner performs a prior art search, and are presented in the form of a special Office Action. A response is often as simple as selecting a group of claims for examination in the application and either withdrawing or canceling the non-elected claims. The rules allow for applicants to argue against a restriction requirement, but given the nearly universal failure of such arguments and the added cost of attorney time, it is standard practice to acquiesce to restriction requirements and reduce the number of claims.

Applicants that want a streamlined and lower-cost examination process should consult with their patent attorney during the drafting stage and determine whether the claim set being filed with the application is likely to be subject to a restriction. Although restriction requirements are not necessarily a bad thing, they do add cost and time to the examination process, which may not be suitable for all patent strategies—especially when a good patent attorney can often selectively avoid restrictions without much effort at the application drafting stage.

LATER STAGES OF EXAMINATION

The back–and-forth between examiner and applicant can theoretically continue indefinitely as long as the applicant is willing to continue paying the patent attorney to analyze the Office Actions and draft responses. Applicants must also pay additional examination fees in a Request for Continued Examination (RCE), typically after receiving two Office Action responses.

Some examinations take a long time because of a poor examiner, a poorly written patent application, or complications caused by prior art that comes up during the process. Unfortunately, applicants that face tough examination may choose to abandon the application after determining that fighting rejections is not worth the cost or that the likelihood of success is too low to warrant further battling. On the other hand, some examinations can go surprisingly quickly and easily. Regardless of whether the process was long and hard or short and simple, the examination ends when the examiner sends out a Notice of Allowance indicating that all patent claims are in condition for allowance. At this point, paying a final issue fee and possibly filing some final paperwork are the last small hurdles to the application becoming an issued patent. Once the issue fee is paid and all other formal requirements are met, the USPTO will send out an Issue Notification that provides a date on which the application will issue, along with the official patent number.

For some applicants, having a single patent application issue is the end of the patent process. For others, more advanced strategies allow for additional related patent applications to be filed before the patent issues. In other cases, a single patent application can be expanded to multiple applications at the time of filing, before the examination begins, or even in the middle of the examination process. Other strategies provide the flexibility to get a patent that is crafted for an attack against a specific competitor or product. With the basics of the patent process as a foundation, the following chapter provides some examples of advanced patent strategies that can be used to create valuable patent portfolios.

CHAPTER 12 SUMMARY

- The examination process begins approximately one to three years after a nonprovisional patent application is filed.
- Once the examination begins, the assigned patent examiner will conduct a prior art search and determine whether the pending patent claims are new and nonobvious in view of prior art discovered during the search.
- The examiner then issues an Office Action that details rejections of the application—such rejections are a natural and expected part of the examination process.
- The most common rejections are novelty and obviousness rejections.
- Applicants have a limited time to respond to an Office Action by having their patent attorney amend the pending claims and/or argue against the rejections.
- In some cases, there can be many successive rounds of Office Actions and responses between the patent attorney and examiner.
- If the examiner is convinced that the pending claims are new and nonobvious in view of the prior art, the examiner will allow the case and the patent application will then issue as a patent.

13

Advanced Patent Strategies

*"You have to learn the rules of the game. And then
you have to play better than anyone else."*

—Albert Einstein

Having a patent portfolio with multiple assets can serve many purposes. For companies that constantly develop new technology, a series of patents is necessary to protect continuing innovation. However, as discussed in previous chapters, it is not uncommon for a single product to be the subject of many patents instead of just one. The main benefit of having a family of related patent applications is maintaining a priority date that stems back to the earliest patent application in the patent family. A provisional patent application followed-up by a nonprovisional patent application is an example of a simple patent family because the nonprovisional application priority date extends back to the filing date of the provisional patent application. If additional nonprovisional patent applications are added to this patent family, these new applications will also enjoy the early priority date established by the provisional application, even if they are filed years later.

This chapter illustrates how patent families can be structured in various ways, and how a strategy is determined based on many different factors, including the type of technology, the business plan of the company, the stage of development of the business or product, the available patent budget, and the need to attract investors or other business partners. In other words, patent strategies must be tailored to the specific needs of each company.

Applicants that are new to the patent process often choose a simple patent strategy mainly because they are unaware of the options that are available and are operating under the common misconception that individual products are eligible for only one patent or that having multiple patents is prohibitively expensive. Moreover, even when options are generally known, applicants and patent attorneys alike are often not aware of the numerous benefits that a patent portfolio provides aside from providing the right to exclude others from making, using or offering to sell patented technology. The following sections illustrate the range of options for growing a patent portfolio, and provide some examples of how these options are applied to create some of the more common patent strategies.

CONTINUATION APPLICATIONS

Besides claiming priority to a provisional patent application, nonprovisional patent applications can also claim priority to other related nonprovisional patent applications with earlier filing dates. In other words, subsequent nonprovisional applications can receive the priority benefit of earlier filing dates established by related nonprovisional applications. Such later-filed applications, generally called continuation or continuing applications, take on three main forms—plain continuation, divisional, and continuation-in-part. Aside from being related to one or more earlier-filed nonprovisional applications, continuations are just like a normal nonprovisional application and will receive a new filing number and will issue as a patent that is completely separate with a unique patent number.

As discussed in previous chapters, patent claims create exclusivity over only a small portion of the subject matter that is disclosed in the specification and drawings. Continuation applications make it possible to protect elements of the specification and drawings that were not protected in an original application, to protect elements more broadly, or to protect elements in alternative beneficial ways. The primary advantage of a continuation is that these protectable elements maintain the priority date of the parent application so the continuation has priority over the same prior art that the parent application did, even though the continuation is filed at a later date.

The ability to file continuations is not unlimited, however; they must be filed while the parent application is still pending. Accordingly, best practices dictate that continuation applications be filed before issue fees are paid in an allowed parent application or before a parent application goes abandoned.[1]

The simplest type of continuation is a plain continuation, which uses the same specification and drawings as the parent application but introduces new patent claims. These new continuation claims can be of different scope to protect different elements of the specification and drawings or can be used to introduce broader claims that include similar, but fewer, elements compared to the parent application.

For example, as shown in the Venn diagrams in Figure 13.1, the specification and drawings of both the parent application and the continuation have the same scope, but the claims have a different scope. The continuation claims are shown generally including elements similar to those in the parent application, but instead include fewer elements or are otherwise written to make the claims broader. Recall that less is more with patent claims, and that having fewer elements makes

---

1. It is possible to file continuations up until the day the parent application patent issues, but because issuance occurs shortly after issue fees are paid, standard practice is to file continuations at the same time issue fees are paid. It is also possible to file continuation applications within two years of when the parent application issues if a reissue is filed. However, this is not a preferred strategy unless absolutely necessary.

a claim broader and therefore easier for similar products to infringe. In other words, the claims cover more variations of an invention and are therefore more valuable because competitors have a harder time designing around the patent.

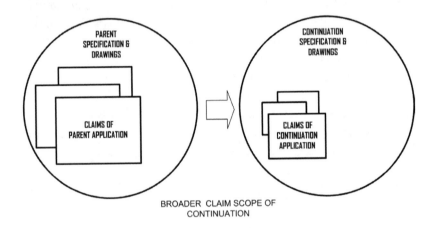

FIGURE 13.1

Alternatively, as shown in the Venn diagram in Figure 13.2, the specification and drawings of both the parent application and continuation are of the same scope, but the claims of the continuation cover different parts of the specification and drawings compared to the parent application. As discussed in previous chapters, a specification and drawings of a patent application typically describe many distinct inventions, yet USPTO rules allow only one invention to be examined per patent application. For this reason, additional continuation applications may be necessary if protection for more than one invention is desired. In many cases, these alternative inventions are just as valuable as the invention claimed in a parent application, or might become even more valuable, depending on how the market develops.

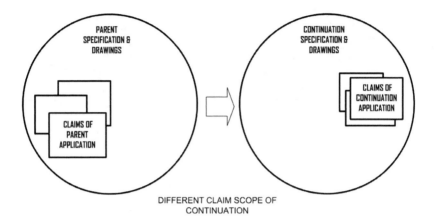

DIFFERENT CLAIM SCOPE OF
CONTINUATION

FIGURE 13.2

Continuations that protect alternative inventions described in an application can be plain continuations, or might be called divisional applications in some contexts. For example, where an examiner issues a restriction requirement and patent claims directed to separate and distinct inventions are required to be canceled from an application, a continuation that includes these canceled claims from the parent would be called a divisional. In many ways, the distinction between a divisional and a plain continuation is irrelevant given that these types of continuation applications effectively have the same logical relation.

On the other hand, a continuation-in-part application (CIP) has important differences compared to divisional and plain continuations that are important to understand. Unlike other types of continuation applications, CIPs are allowed to add new subject matter to the specification and drawings, in addition to having new claims.

For example, in the Venn diagram figures in Figure 13.3, the CIP application has expanded the scope of the specification and drawings to include new matter as indicated by the additional sliver added onto the circle. This invention disclosure, or new set of elements, becomes available for use in the claims, and the new claims of the continuation include elements that are within the scope of the new matter presented in the specification and/or drawings. Additionally, elements from the old specification and drawings are also present in the new claims.

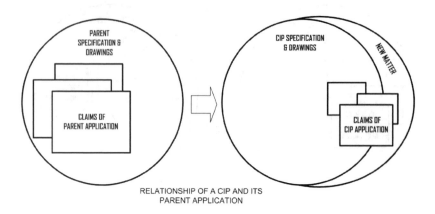

RELATIONSHIP OF A CIP AND ITS
PARENT APPLICATION

FIGURE 13.3

However, the downside of CIPs is that claims that include new subject matter are only entitled to a priority date of when the CIP was filed and not back to the earliest priority date of the parent application like other continuation applications. Moreover, the specification and drawings of the parent application can be used as prior art against the claims in a CIP. On the other hand, claims that only include subject matter from the parent application still retain the earliest priority date benefit. Given the limitations of CIPs, they are primarily used only when important changes to an invention occur after a nonprovisional application has already been filed, and these changes warrant protection, but not a new stand-alone patent application. In other words, CIPs are typically used as a method of "updating" nonprovisional patent applications.

With CIPs being applicable in limited circumstances, and given the choice between using a plain continuation and a divisional based on whether there is a restriction requirement in the parent application, applicants are effectively only able to selectively implement continuations of the type that do not add new matter to the specification and drawings. With this in mind, the following patent strategies refer to continuations in general and assume that divisionals or CIPs will be used where applicable or necessary.

EXAMPLES OF CLAIM/PORTFOLIO STRATEGIES

With a variety of options available for building a patent portfolio, applicants are best served by selecting a patent strategy that fits their unique business goals and budget. However, the attorney time it takes to fully describe all the options and collaboratively craft a tailored plan is often not within the budget of many companies, or applicants may not even realize that different options are available. Instead, patent attorneys will either apply a one-size-fits-all plan or independently select a strategy based on limited knowledge of the client. Unfortunately, by not being an active part of the patent-planning process, applicants miss out on a critical opportunity to not only pick a plan that fits their business, but also to adapt their business plan based on the opportunities that having a patent portfolio provides. Accordingly, the following should provide inspiration for the types of strategies that can be applied to existing business plans and ways that patent assets can add previously unknown dimensions to a company.

THE "BASIC"

The most common patent strategy is filing a single, nonprovisional patent application (possibly after filing a provisional patent application first) with a set of patent claims that includes no more than three independent claims and no more than 20 claims total. These limitations come from USPTO rules that impose additional fees for claims above this range, and the fact that exceeding this range makes it more likely that the claims will be restricted to a smaller claim set. In the Basic strategy, claims are written with a goal of being broad enough to spark some rejections by the examiner, but with the aim of wrapping up the examination process within two or three Office Actions. Just before the patent issues, applicants might choose to file a continuation application if there are important inventions that were not protected in the original application.

This strategy can be a good fit for an average company where patent protection is not a major component of the company's business plan and marketing strategy. This basic strategy requires an average patent budget and leaves open the possibility for expanding the patent family down the road if need be. It is intended to obtain a reasonably broad patent within a reasonable amount of time.

THE "WEDGE"

In an alternative strategy, applicants may be able to receive a pair of patents on the same invention with the first one issuing faster than the Basic plan and the total cost being about the same or only slightly more. This strategy begins by drafting a narrow set of claims (long claims with many elements) in a first nonprovisional application that is intended to be allowable without an Office Action or with a goal of having examination complete within one Office Action. Before this first application issues, a continuation is filed that uses experience from the parent application to draft a set of broader patent claims. The new claims of this second application are drafted more aggressively but with the aim of completing examination within one to three Office Actions.

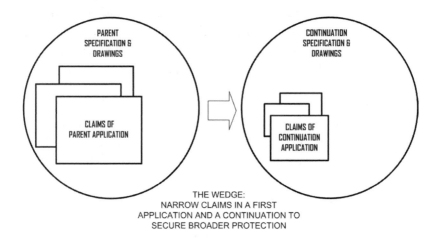

THE WEDGE:
NARROW CLAIMS IN A FIRST
APPLICATION AND A CONTINUATION TO
SECURE BROADER PROTECTION

FIGURE 13.4

This strategy has several benefits over the Basic plan. For example, applicants may get an issued patent faster, which translates into reduced cost in the examination phase of the first application. Although the claims of the first application are not as broad as they could be, a continuation is filed that goes after the broadest protection available, and has the benefit of experience from the parent application to hopefully expedite examination. Just as the tip of a chisel leads the way for its wider upper shaft, the Wedge technique refers to easily getting narrow claims allowed in a first application and using knowledge and rapport with the examiner from the first application to get much broader protection in a subsequent continuation application. The total cost of these two patents can be about the same as or a bit more than a single patent in the Basic plan, but results in a portfolio of two patents instead of one.

Additionally, because the first patent issues faster and with less cost, it is a good option for companies that have tight patent budgets. For example, the Wedge allows a company to obtain a narrow patent quickly and preserves the option of pursuing broader claims in a continuation. In other words, the continuation step is optional, and if applicants are not able to invest additional capital in a continuation, they can settle for the single narrow patent instead.

The Wedge strategy is also a good fit for companies that will receive positive marketing value from having an issued patent faster. For example, companies are able to promote having an issued patent (and a continuation application) relatively quickly, which may be more attractive to potential investors and business partners. Additionally, having an issued patent may act as a deterrent to competitors, scaring them off before they get established.

Conversely, the issued patent gained quickly by the Wedge strategy will likely have weaker enforcement value because it is so narrow. In other words, the compromise of avoiding a lengthy and expensive examination is that the patent may have claims so narrow, that it is extremely unlikely that competitors will actually infringe them. However, the claims are unlikely to receive such strong scrutiny unless the patent is actively enforced or the subject of licensing or sale. This is where the continuation in the Wedge strategy becomes important. Although the first patent is narrow, the continuation provides the opportunity to capture the broadest claim scope possible, which makes it more likely to be infringed and, thus, more valuable.

THE "PORTFOLIO SPLIT"

For purposes of claim diversity, it is common practice for patent attorneys to draft nonprovisional applications that have two or more independent claims, with respective sets of dependent claims associated with each independent claim.[1] For example, for a smartphone app, it is possible to draft independent claims directed to method steps performed at the user device, method steps performed at a server, a user device that performs a certain method, and a server system that performs a certain method. Other independent claims might be directed to method steps performed by both the user device and server, and directed to a computing system that includes user devices and a server that perform certain steps. Other types of inventions, such as mechanical products, can also be the subject of a diversity of independent claims, including claims directed to the product directly, a method of using the product, and a method of manufacturing the product.

In strategies like the Basic, these separate independent claims are included in the same patent application. Sometimes an examiner might subject the claims to a restriction requirement and require selection of one independent and its dependent claims, but often many different independent claims can be examined and issue together in the same patent.

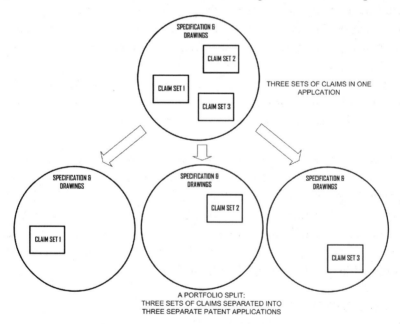

FIGURE 13.5

---

1. Independent and dependent claims are discussed in detail in Chapter 12.

However, applicants have another option for these different independent claims. Instead of presenting them in a single application, each separate independent claim, and the claims that depend from it, can be split up into separate patent applications. This simple change in the structure immediately catapults a company from having a single patent application to having a patent portfolio with two or more patent applications.

This can be immensely valuable given that a patent portfolio's value is often considered to be directly proportionate to the number of patents or patent applications, and not the number of claims or even the content of the application. It is an extremely costly and subjective task to put a true valuation on patent assets, so general valuations are typically based on quantity and not content of patent assets. For example, where patent assets are each considered to justify an increased company valuation of $1 million, splitting a single patent application into three patent applications could generate an additional $2 million dollars in value.

Luckily, splitting a patent application is extremely inexpensive compared to the value it creates. For example, once the specification, drawings and claims for a first patent application are drafted, splitting them is only a matter of formatting, and ideally some superficial changes to the title, abstract, and possibly the background section. A nearly identical specification and drawing are used in each new application, with the original claims split among the applications, and minor changes to make the applications appear different. Up-front additional cost is therefore only a small amount of attorney time in splitting the original application and additional USPTO filing fees and filing time for each extra application. This incremental additional cost is negligible compared to the potential for doubling, tripling or even quadrupling the value of a patent portfolio depending on how many applications are filed.

The downside to the Portfolio Split is that multiple applications incur more cost once these applications come up for examination. For example, if three applications are filed instead of one, there will be three parallel examinations in approximately one to three years. In some cases, lessons learned in the first examined application can expedite examination in other applications, but it is best to assume that having three applications might triple the examination cost.

The Portfolio Split is therefore a great option for companies where the patent portfolio value is important. Start-ups hunting for investors, business partners, or even positioning themselves for acquisition tend to be great candidates for such a strategy. Additionally, companies making an initial public offering (IPO) of stock often use such a strategy to increase their stock price leading up to the sale and afterward. Additionally, if marketed correctly, a larger patent portfolio can increase goodwill with consumers and scare competitors. Given that patents are an indicator of innovation, having more patents is viewed as being more innovative and more protected.

On the other hand, the Portfolio Split may not be the best option for companies where the patent portfolio value is less important to a company's overall business strategy. Additionally, the increased cost due to multiple patent examinations is a major downside for companies with limited capital. Ideally, building a patent portfolio with the Portfolio Split will attract more working capital through investments, partnerships, acquisition, selling securities, or improved sales, but the risk of having to support several patent examinations might be too much for some companies.

THE "STAGGERED PORTFOLIO SPLIT"

In a typical Portfolio Split, all of the split applications are filed on the same day. In contrast, a Staggered Portfolio Split files the split applications in succession, often separated by several months. In this strategy, the first application is a parent application; subsequent applications are continuations from the first application and therefore enjoy the same early priority date. This staggering does not result in any additional cost compared to the cost of filing the applications all at once.

The purpose of the Staggered Portfolio Split is to create a different marketing impact. Companies often issue press releases to the public and shareholders announcing the filing, publication, or issuance of patent applications, which often provide a bump in stock prices and otherwise improves the image or valuation of the company. Announcing patent activity is typically an outward signal that a company is continuing to be innovative, building exclusivity in the marketplace, and increasing its overall value.

Under the right circumstances, a Staggered Portfolio Split has substantially more impact than a single announcement that several patents were filed at once. However, this strategy might not be suitable for small companies with a limited media presence, companies that already regularly file patents, or where announcing patent activity otherwise has limited benefit. On the other hand, a Staggered Portfolio Split can be a good option for companies that are ramping up to an IPO or for publicly traded companies that need a series of positive press releases related to patents.

THE "KITCHEN SINK"

The strategies discussed above relate to cases in which applicants have a single product or invention and have the choice of filing a single patent application or splitting it into multiple applications. Closely related variations can be included in the same application, but where variations are extremely different or where completely separate products or inventions have been developed, it is typically necessary to have completely separate patent applications for adequate protection. Patent examiners might allow claims to small variations of an invention, but will definitely make a restriction requirement for claims directed to completely different products or inventions. Accordingly, protecting separate products or inventions will require separate patent applications.

However, there is no limit to what can be described in the patent specification and drawings because restriction requirements apply only to the patent claims. Theoretically, applicants could file a single patent application that simultaneously discloses a cure for cancer, a new smartphone design, a novel wine opener and a search engine that beats Google. In certain limited cases, it can be cost-effective to describe multiple inventions or products in a single application.

For example, where applicants have developed several completely different variations of a product but have not yet decided which one is the most commercially viable, having a single omnibus application will cost substantially less than several parallel applications and provides the option to pursue claims directed to a single preferred product later on. This strategy gives applicants at least until examination begins to do market testing and further develop the product before selecting the best version. Even if a final decision is not made by the time the application comes up for examination, filing continuations allows all variations to remain active for claiming at a later time. Similarly, where inventors have several great ideas, but know that only one will be the first subject of a business, a patent application that describes all of the ideas can be filed, and once the business direction is solidified, the selected idea can become the focus of the patent application at a later date.

This Kitchen Sink strategy is great for inventors with a limited budget that have many ideas or product variations, but need time to determine which ones will be the subject of an ongoing business. If there is an insufficient budget to support examination of several individual applications to each different idea, and having a portfolio of several applications is not beneficial, having a single application might be a best option.

Additionally, this strategy allows early inventions to remain actively available for protection while also protecting new alternative inventions that come along during the research and development process. For example, continuation-in-part applications can be used to add new matter while also keeping earlier subject matter alive as long as at least one continuation or an original application is still active.

THE "LITIGATION LASER"

Most patents are like bear traps. They are designed to lie in wait until someone comes along and inadvertently or unknowingly wanders along with a product that infringes the claims of the patent. At the time patent applications are filed, applicants sometimes have a commercial product in its early stages or may not even have a prototype built at all. Direct copiers or competitors with similar products may not even come into existence until years after an application is filed and often not until after a first patent application has already issued.

This creates a difficult task for patent attorneys who are expected to craft patent claims that still cover theoretical and unknown competitor products that might emerge many years later. Moreover, once a patent issues, competitors can read and analyze the patent claims to develop a product that is extremely similar, but does not infringe because it does not have all of the elements defined by the issued claims. Luckily, continuation applications provide a solution to this problem and allow for the Litigation Laser technique.

For example, suppose that a competitor comes out with a product that is dangerously close to but does not infringe the patent claims of an issued application. Unfortunately, there is little recourse with this patent if the product does not have all of the elements of the patent claims. However, if there is an active continuation in this same family, new patent claims can be drafted that can directly read on this product, assuming there is sufficient support in the existing patent specification or drawings. (Adding new matter in the specification and drawings would be a CIP and would give resulting claims a priority date after the competing product came out.) Unlike the original claims that had to predict what future products might be like, the patent attorney now has the luxury of having the competitor product sitting on his desk while drafting patent claims that directly read on it.

This strategy is only available, however, if there is an active application in the patent family that has a priority date that precedes the new product. As a result, many companies will file continuation applications before the last active application in a patent family either issues or goes abandoned so that the option to specifically attack competitors remains open indefinitely. In patent families where continuations are already part of a patent strategy, the Litigation Laser is just another benefit of having an active continuation. For other cases, keeping this option open may be a primary motivation to file continuations aside from broadening protection or protecting alternative embodiments.

A good example of a similar strategy at work is in a case that was tried by a legendary Seattle patent litigator representing plaintiffs against an alleged infringer. In the early stages of the case, the defendants demonstrated that their product did not infringe because certain elements recited in the claims were not present in their device. However, this crafty patent litigator was able to get the defendants to make statements about what specific elements *would* be infringed if they were present in a set of claims. Armed with this knowledge, he quietly modified a pending continuation to include claims that would certainly be infringed based on the defendants' admissions.

Given that patent claims are not enforceable until they issue, he had to wait until the application was examined and allowed. Luckily, the continuation came up for examination quickly and his dead-to-rights claims were allowed by the examiner as the case was in progress. Knowing that the defendants were unaware of the allowed continuation, he walked them through a "theoretical" set of claims that would be infringed based on their previous statements. With these admissions in hand, he confronted the defendants with the claims of the allowed continuation. The defendants settled that same day.

SOME FINAL THOUGHTS

The example strategies discussed above should illustrate that patent protection is not a one-size-fits-all endeavor. This is especially true for companies with fewer patent assets or a limited patent budget, and where there is substantially more pressure for each patent or application to provide maximum benefit for its cost. Applicants must therefore work closely with their patent attorney to craft a patent strategy that not only best suits the business goals of a company, but that also works within current and future budget constraints. Regardless of the chosen strategy, applicants must actively exploit their patent assets in order to derive the most value from them. In addition to the numerous benefits gained by patent protection that have already been discussed, the following chapter further explores how to leverage patent assets while they are still pending and after they become issued and enforceable patents.

CHAPTER 13 SUMMARY

- Continuation applications are nonprovisional patent applications that are related to and claim the priority date of another nonprovisional application.
- Continuations can be used to pursue broader claims to an invention; to pursue claims to different aspects of an invention; to expand a patent portfolio; or to add invention updates to a nonprovisional application.
- Applicants can often choose to split a single application into more than one application.
- In some cases, it can be helpful to file a single application that describes many different inventions, which would otherwise be filed in several separate applications.
- If the original specification or drawings have sufficient description, applicants can use a pending application to present patent claims that directly read on a new competitor product.

14

Leveraging Patent Assets

"If people don't get paid for their inventions, that's not a good thing. In the case of many patents, there are people who aren't in a position to take them to the next level. If you don't enforce your rights, no one is going to enforce them for you."

—Nathan Myhrvold, founder of Intellectual Ventures

"IBM isn't investing billions of dollars every year into research and development— and winning more patents than our top 10 competitors combined for more than a decade—as an academic exercise."

—Samuel J. Palmisano, former President and CEO of IBM.

After patent applications are filed, while they are pending, and after they issue as patents, applicants will want to leverage these patent assets for their benefit. Issued patents confer the right to exclude others from making, using, selling, offering to sell, and importing any product or activity that is defined by the claims of the issued patent. This exclusionary right can be used in a direct way for stopping competitor products or methods. Alternatively, selected parties can be allowed to sell or use a patented invention in exchange for a cash payment or other consideration. Such a license is effectively an agreement not to enforce the patent against these selected parties. Issued and pending patent assets can also be sold outright or even used as collateral for a loan. Aside from the direct monetary or exclusionary benefits that patents provide, the marketing and public relations benefit of patent assets is also important, yet often overlooked or under-utilized. With a comprehensive understanding of the patent process as a foundation, this chapter goes into more detail about the many ways that patent assets can be leveraged.

ENFORCING A PATENT

Patent rights can only be enforced once a patent issues. This should make sense given that the claims in a pending application have either not been examined or have not yet been approved by a patent examiner. The final claims could be narrow, broad, or might never issue as a patent for a variety of reasons. Asserting "patent pending" status can still be a strong deterrent to competitors, but nonetheless has no teeth until the day the patent application officially issues as a patent.

When competitors begin to enter the market with similar products while a patent application is still pending, applicants tend to get upset and want to take immediate action to block competing products or services. Finding out that no action can be taken until the patent issues, their first inclination is to try to get a patent issued as soon as possible so that they can attack. However, enforcing patents is a delicate process that is best done slowly and deliberately. Instead of rushing to attack competitors, a better strategy is to quietly gather information about competitors and to go on the offensive when the timing is right.

For example, as discussed in the previous chapter, if applicants have a live application in a patent family, patent claims can be custom-tailored to competitor products. By taking the time to gather information and to clearly understand the elements of any allegedly infringing products, a stronger case for infringement can be built. Moreover, waiting to enforce patents within a reasonable time period does not have a downside because damages continue to accrue during the time of infringement and can even extend back to when a pending patent application was published.1 Accordingly, enforcing patents should include a carefully crafted plan orchestrated by patent attorneys and litigators.

One common misconception is that enforcing patents is a simple process of sending letters to infringers, who will immediately recognize their infringement and either cease selling or using the patented invention, or will sign up to pay an ongoing royalty for use of the invention. Unfortunately, enforcing patents is typically not that easy and sending out such "cease-and-desist" or "invitation-to-license" letters is often pointless and might unintentionally land patent owners in a world of trouble.

For example, the most common response to such letters is to simply ignore them as idle threats. The rationale for this comes from the fact that filing and pursuing a patent infringement lawsuit is an expensive and time-consuming endeavor that most patent owners are not truly willing to undertake. Accordingly, the assumption is that the vast majority of patent owners who send such letters will never follow through on enforcement, even if there is infringement.

Some attorneys will respond to these letters with a general response that their clients do not infringe and possibly an allegation that the patent at issue is invalid or otherwise unenforceable.

On the other hand, another response is filing a declaratory judgment (DJ) lawsuit against the patent owner that sent the threatening letter. Such an action is a lawsuit effectively requesting that a court officially declare that the party is not infringing the patent at issue. DJ actions can be filed in response to any letter that contains the threat of patent litigation, and current law considers even the friendliest letters that mention patents as a veiled threat to sue. Being drawn into a DJ

---

1. *See* 35 U.S.C. 154(d)

action is as serious as any other type of lawsuit and requires hiring an attorney to respond to the allegations.

In addition to being at a distinct disadvantage by being put on the defensive and not being able to control the timing of the lawsuit, the party that files a DJ action can choose the venue in which the case will be tried, which can add cost and create additional disadvantages in an already difficult situation. For example, suppose a small company based in San Francisco sends a cease-and-desist letter to a company based in New York alleging patent infringement by widgets being sold in San Francisco and other markets. The New York company could file a DJ action in New York and the San Francisco company would be forced to defend itself there. Accordingly, cease-and-desist or invitation-to-license letters are typically weak strategies, because, while many companies will shrug them off, others will drag the patent holder into an expensive and undesirable DJ action.

Such letters are therefore only used with extreme caution and in extremely limited circumstances. For example, when the target is an individual or small company that is unsophisticated with patents and unlikely to seek out counsel of a patent attorney, sending such letters might be effective in scaring the company into taking a license or stopping sales or use of a patented invention. Additionally, where the patent owner has built a reputation for following through on litigation threats, such letters might be taken more seriously. Regardless, patent owners should never send out letters to competitors regarding patents without first consulting a patent attorney and carefully weighing the pros and cons of such an approach.

The best way to be taken seriously and open up negotiations is to actually file a patent infringement lawsuit. Even the largest company will be required to respond to such a lawsuit and will be forced to make a determination of whether the patent at issue is actually being infringed. However, filing an infringement lawsuit is not a trivial undertaking and patent owners should be willing and able to see the case through to trial if the alleged infringer decides to fight back. Bluffing can be dangerous when facing a well-capitalized opponent that might try to outspend you instead of settling the case. Again, such a strategy should be carefully planned and executed by a trusted patent litigation attorney.

Regardless of the chosen strategy for enforcing patents, the first step is to determine if a patent is being infringed and to identify the strengths and weaknesses of arguments for and against infringement. A patent attorney or patent litigation attorney should be consulted in making a final determination and doing a comprehensive analysis, but patent owners will benefit from doing a preliminary analysis before seeking expert counsel. Such a preliminary screening for patent infringement is very similar to the process of analyzing novelty rejections in an Office Action as discussed in Chapter 12, and begins with creating a claim chart that includes all of the issued claims separated into constituent elements.

For example, using the sample claim for a bicycle invention discussed in Chapter 11, the elements are split into individual cells in a first column of a table, and analysis of each element is provided in the second column. This example only includes one independent claim, but a full analysis should cover all of the issued patent claims including both independent and dependent claims.

Patent Claims	Alleged Product
1. A vehicle comprising:	Yes, the bicycle is a vehicle.
a front and rear wheel;	Yes.
a frame including handlebars and	Yes.

at least two bars spanning between the wheels;	Yes.
a petal assembly coupled with the rear wheel via a chain;	Yes.
a pair of brake levers disposed on a set of handlebars.	No. Brake levers are not present on the handlebars.

For each claim in the analysis column, indicate whether the element is present in the accused device, and provide some explanation of why the element is or is not present. One good way is to provide a picture of the accused device and highlight the element as shown. Where a picture is not applicable, present proof of why the element is present or indicate that it is not present.

In some cases, further investigation might be required to determine if a given element is present in an accused product. Here, it is fine to indicate that this part is unknown and describe what analysis would be necessary. Also, it is advisable to describe any assumptions that are being made and indicate where claim language is unclear or difficult to understand.

To prove infringement, all elements of at least one claim in the patent must be satisfied by the accused product to prove infringement. If even one element is missing from any given claim, then that claim will not be infringed. Additionally, if an independent claim is not infringed, then all of the claims that depend from it will also not be infringed because the elements of the independent claim are also part of these dependent claims. In the example claim chart above, all elements are satisfied up until the last one because there are no brake levers shown on the accused product. Unfortunately, this indicates that the product as shown would not likely infringe this patent claim. This shows how important it is to analyze each and every element of a patent claim because even one element missing or arguably not present can be fatal to a case of infringement.

Since claim charts like this one will be required when formally building a patent litigation case, a preliminary applicant assessment can be helpful in identifying where more research and analysis of a competitor product or service is necessary, or is helpful in recognizing that a similar product is not actually infringing because of a few missing key elements. Regardless, a patent litigation attorney should review any analysis before conclusions are reached about whether infringement is present.

LICENSING, SALES, AND LOANS

As discussed in Chapter 2, patent assets can be licensed, either exclusively or nonexclusively to other parties for ongoing royalties, upfront payments, and any other sort of consideration. Additionally, patent assets can be used as collateral for loans or can be sold.

Filing a patent infringement lawsuit, sending out offer letters, or sending out cease-and-desist letters as described above can lead to a settlement that includes a license or might even lead to the patent assets or company as a whole being willingly sold to the allegedly infringing competitor. However, there are other ways of finding buyers and licensors for patent assets that are much less adversarial. For example, networking with companies that use or are interested in similar technology can lead to collaborative business relationships where such companies will adopt patented technology and pay a reasonable royalty to use the invention or even buy the patents outright. Ideally, patent owners already have connections within an industry that give them a foot in the door at potentially interested companies. If not, patent owners will need to work to establish these relationships themselves or partner with industry insiders who have the right connections.

Another option for selling patent assets is using a patent broker to secure interested buyers. Patent brokers will take a percentage of the total amount of the sales price for a patent portfolio; a commission that can be well worth it if the broker is able to find a buyer where sales attempts have been unsuccessful. When seeking a patent broker, it is wise to shop around. Create a list of approximately 20 patent brokers by doing an Internet search. After research on each, separate the brokers into about four groups of five, from most to least desirable. Some ranking factors might include technology specialties, size, and number of recent deals, and the apparent professionalism of each broker. Start by contacting the brokers in the top-tier group first. If no serious interest materializes within a week or two, move on to the next group.

Just as with finding a patent attorney, you want to find a broker that is legitimately excited to work with you and your patent assets, and understands the technology well enough to appreciate its value. By contacting several brokers at once, you are more likely to find one that is a good fit for selling your patents. For example, you might find that some brokers already have specific buyers in mind, whereas others might not see much promise in taking you on as a client. Often this has less to do with the patents and more to do with the broker's set of industry and buyer connections or their capacity to take on new clients.

Similarly, when looking for a bank or other lending institution to make a loan based on patent assets, it helps to shop around and find a lender that is a good fit. Lenders that take a lien on patents as collateral for a loan will put a value on the assets that directly impacts the loan amount, interest rate, and lending terms. Given that patent valuation is extremely subjective, lenders that do not understand the technology or the value of the technology will give a low valuation that will likely result in less desirable loan terms. Finding a lender that is enthusiastic about the patents used as collateral and that can also offer good terms is therefore of utmost importance.

MARKETING RELATED TO PATENTS

Marketing efforts and patents assets should work in synergy. As discussed in previous chapters, all marketing efforts related to patentable inventions should ideally be postponed until at least a provisional patent application is filed. At the very least, U.S. patent applications must be filed within one year of a first public use, public disclosure, or offer-for-sale. However, once the first patent application is filed, a company should not hold back on marketing efforts, at least for inventions that are disclosed in a patent application. For updates, improvements, and changes to an invention, a follow-up application may need to be filed before these new aspects can be safely disclosed.

KNOW YOUR NUMBERS

There are three main numbers related to patent cases—an application number, a publication number, and a patent number. Every patent application receives an application number that has eight digits, which are typically formatted as 12/345,678. When nonprovisional patent applications publish, this publication receives a ten-digit identifier that has a year followed by seven digits and is often formatted as 2000/1234567. When utility patents issue, they are sequentially numbered. Issued utility patent numbers are currently into high seven figures and are often formatted as 1,234,567.

Marking products as "patent pending" or "patented" is an important marketing tool because it fosters goodwill with the public, while also warning competitors that attempts to copy the product or make something similar could make them liable for damages. For products, packaging, and other locations near an inventive product, markings of "patent pending" should be applied after filing a patent application that describes the invention, regardless of whether it is a provisional or nonprovisional application. However, the "patent pending" marking should be removed if a provisional patent application expires without a follow-up nonprovisional application being filed, or if a pending non-provisional goes abandoned during the examination process.

This "patent pending" marking should be replaced with a "patented" marking once a patent application issues as a patent. To avoid liability issues related to patent mismarking, a safe designation is something along the lines of "Product may be covered by one or more of the following patents," followed by a list of issued patent numbers. Alternatively, recent changes to patent laws provide for "virtual" patent marking[1] that allows for displaying an Internet URL instead of a listing of specific patent numbers. Accordingly, a good virtual marking would be "Product may be covered by one or more of the patents listed at www.CompanyName.com/patents."

In addition to proper marking of products that are the subject of patents or patent applications, companies can also benefit from publicizing the important milestones of the patent process, including filing of applications, publication of a nonprovisional application, and providing notice of when a nonprovisional patent application issues as a patent. Given that some people mistakenly view provisional patent applications as being inferior to nonprovisional applications, it is typically best to simply state that a patent application was filed. An application number should not be provided because this information reveals that it is a provisional or nonprovisional application. Moreover, the application number does not provide people with any useful information because it is not possible to look up the provisional application and a nonprovisional will not be publicly available using an application number until it publishes. Publishing an unsearchable number

---

1. 35 U.S.C. 287(a)

can leave a bad impression because some might believe that a patent application was not really filed and might leave searchers frustrated.

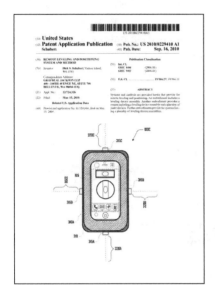 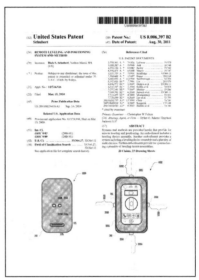

FIGURE 14.6   Example of a patent application publication (left) and a publication of the same application once it issued as a patent (right)

Once a nonprovisional patent application publishes, applicants should consider including the publication number in any announcement and possibly a link to the formal publication. This publicly available publication of the pending application has the same formatting as a publication of an issued patent, aside from having a publication number instead of a patent number. Plus, the formal USPTO formatting of the document can be extremely impressive and creates an air of importance for the invention and applicant. Similarly, once a patent issues, announcements can include the patent number and a link to a copy of the official USPTO publication of the patent. These important milestones in the patent process can be the subject of formal press releases, blogs, social media posts, or even an email to a contact list.

CHAPTER 14 SUMMARY

- Patent rights can only be enforced once a patent application issues as a granted patent.
- Carelessly sending cease-and-desist letters is a dangerous practice and can result in a declaratory judgment lawsuit against the patent owner.
- Claim charts can be drafted to determine if a given product might infringe an issued patent.
- Patents and patent applications can be monetized through licensing, sales, or use as loan collateral.
- Strategically publicizing pending patent applications and issued patents can have marketing benefits for a company and also deter competitors.

Epilogue

I hope this book inspires you. Regardless of whether you are planning to invent a revolutionary new product, already have a developed prototype, or have a patent application on file, I hope this book has expanded your view of how patents can be used to add significant value to your company. Moreover, I hope this book has given you a roadmap for both the immediate and long-term steps necessary for adequately protecting your valuable ideas and inventions.

Much of the information of this book is known only by seasoned patent attorneys or people with years of experience with patents. The information provided can help you maximize the potential of your intellectual property and avoid common patent mistakes. Unfortunately, many inventors and entrepreneurs that are new to the patent process learn the pitfalls described in this book the hard way, unknowingly forfeiting or substantially compromising their patent rights. While some are able to recover and avoid future mistakes, others never get a second chance. Some issues might become apparent early in the life of a business, whereas others can lie hidden for years.

As you now know, inventions should ideally be held in secret until a patent application is filed. Patent rights in nearly all foreign countries are immediately lost upon a first public use, public disclosure, or offers-for-sale, and U.S. patent rights are lost one year after such activities. Filing a patent application as soon as possible is also important because the United States now gives priority to inventors who file first, not to those who invent first. Many inventions can be protected at early stages of development—it is only necessary to have enough detail to allow a patent attorney to draft a patent application that would enable one of ordinary skill in the art to make and use the invention. A working prototype is not required. In fact, in the eyes of the law, filing a patent application is equivalent to having a working prototype.

You also now know how important it is to file patent applications that adequately and broadly describe an invention. Although the USPTO will accept essentially any document that is filed as a provisional patent application, inventors only get priority for what is sufficiently described. Unfortunately, it is far too common for inventors to unknowingly file provisional patent applications that provide absolutely no protection because the technology is not adequately described.

For this reason, among many others, it is essential to engage a trusted patent attorney for assistance in planning and executing a patent strategy. Patent attorneys provide value by ensuring that patent applications broadly and sufficiently describe an invention, and can negotiate with the USPTO to get the broadest patent protection possible during the examination process. The goal of the patent process is not simply to get an issued patent, but is to get an issued patent that broadly protects the invention, while avoiding defects that make it invalid or otherwise unenforceable.

To achieve this goal, you now know that the patent examination process should be viewed as a negotiation. Your patent attorney carefully sets up an initial offer when a nonprovisional application is filed and negotiates with the patent examiner to get the broadest patent protection possible, while also convincing the examiner that the patent claims are new and nonobvious over the prior art. Accordingly, a series of rejections and responses between examiner and attorney should not only be expected during the examination process, but should also be welcomed.

One of the key insights this books provides is that patent claims are the most important part of a patent and are what defines the scope of intellectual property ownership granted by the patent. During examination, the negotiation is over the wording of the claims, and whether this wording describes an invention that is new and nonobvious in view of documents the examiner discovered during a prior art search. When determining whether a patent is infringed, the question is whether an accused product has all elements recited in at least one patent claim. For this reason, patent insiders know that a basic under-standing of patent claims is critical to understanding patents, even though the claims are strangely written and difficult to parse.

In addition to the patent basics described above, you are also now aware of some advanced patent strategies and techniques that allow companies to build custom-tailored patent portfolios and then exploit them to derive maximum value. For example, in addition to the abil-ity to maintain exclusivity in the marketplace or selectively license the right to practice an invention, patents can play an important role in marketing. Not only do patents provide disincentive for others to enter a market or to create competing products, patents help brand a com-pany and its technology as being superior and innovative. This attracts customers, business partners, and investors alike.

The number of patent assets a company owns is typically viewed as directly proportional to the value and protection afforded by such assets. A company with ten patents instead of one is often viewed as being ten times more innovative, ten times more protected, and having a patent portfolio with ten times more value. However, as discussed throughout the book, another key point is that companies often have a choice of whether a given product is protected by one or several pat-ents. Most inventors do not realize that a single innovation can often be the source of numerous unique patentable inventions, and by filing patent applications that have patent claims with different wording—sometimes only slightly different—it is possible to selectively grow a patent portfolio. In addition to providing actual value by creating alternative ways for others to infringe, having multiple patents has greater perceived value. If done correctly, the cost of having multiple patent assets does not proportionately increase the cost of building such a portfolio. A portfolio with ten patents need not cost ten times as much as a portfolio having only one patent.

The information in this book is intended to give you an overview of the many options available and the issues to be mindful of when developing and patenting inventions, but it is not intended to replace the integral role that a patent attorney should play in formulating and executing a custom intellectual property plan. Each business and invention is unique and thus requires an expertly crafted patent strategy that is specifically suited to a given set of facts, circumstances, and the current state of the law.

Your next step is to take action. If you who want to change the world with a new product, dedicate yourself to inventing it. If you already have a product idea, seek out a patent attorney for an initial consultation, even if you are at early stages, and plan how you are going to bring your product to market while also protecting your intellectual property. For those of you who already have patents and/or patent applications in place, think about how you can derive more value from these assets and whether it would make sense to expand your existing portfolio.

In a field where good advice is often expensive and hard to come by, I hope this book will be a cost-effective resource that acts as a guide throughout the inventing, patenting, and business growth process. For additional content, updated information, and to share your experiences, visit www.PatentsDemystified.com.

Index

N
Nike, 4
Nondisclosure agreement
(NDA), 54–56
Non-practicing entities (NPEs),
27
Nonprovisional patent
application, 41
drafting, 180–184
post-filing considerations,
188–190
publish of, 247
reviewing, 184–186
steps before filing,
186–188
understanding patent
claims, 173–180
working with patent
attorney, 171–190
Nonprovisional utility
application, 88
cost of, 88–89
provisional utility vs.,
88–98
real-world examples, for
choosing, 94–98
value of patent,
technology over time,
92–94
Novel algorithm, 51–52
Novelty rejections, 196–201

O
Obviousness rejections,
201–207
Offers-for-Sale, 52–54

Office action, 42
assisting with, 207–214
deciphering, 193–217
definition of, 194
finding patentable
elements, 211–214
rejections and objections,
214–216
Owens Corning, 5

P
Patentability, 72, 78
vs. infringement, 78–80
to understand, chances
of, 69
Patent agents, 116
Patent application, 32–33, 48
drafting of, 105–106
filed by multiple
inventors, 63–64
prefiling procedures and
considerations,
47–66
prototypes, to fill, 62–63
too early vs. too late,
60–62
Patent Application Information
Retrieval (PAIR) system,
195
Patent assets
enforcing, 238–243
leveraging, 237–238
licensing, sales, and loans,
243–245
marketing related to,
245–247